The Lawful Forest

The Lawful Forest

A Critical History of Property, Protest and Spatial Justice

Cristy Clark and John Page

EDINBURGH
University Press

Edinburgh University Press is one of the leading university presses in the UK. We publish academic books and journals in our selected subject areas across the humanities and social sciences, combining cutting-edge scholarship with high editorial and production values to produce academic works of lasting importance. For more information visit our website: edinburghuniversitypress.com

Edinburgh University Press Ltd
The Tun – Holyrood Road
12(2f) Jackson's Entry
Edinburgh EH8 8PJ

First published in hardback by Edinburgh University Press 2022

Typeset in 11/13pt Adobe Garamond Pro
by Cheshire Typesetting Ltd, Cuddington, Cheshire

A CIP record for this book is available from the British Library

ISBN 978 1 4744 8744 3 (hardback)
ISBN 978 1 4744 8745 0 (paperback)
ISBN 978 1 4744 8746 7 (webready PDF)
ISBN 978 1 4744 8747 4 (epub)

Contents

Acknowledgements

This book has its origins in the 2017 annual conference of the Association for Law, Property & Society (ALPS) at the University of Michigan in Ann Arbor, including the field trip to Detroit. A further trigger was the judgment of *Brown v State of Tasmania,* where the High Court of Australia upheld a (limited) right to protest in Australia partly on the basis of a pre-existing right to access the 'public forest estate'. We were so intrigued by the Court's recognition of this public forest estate that we decided to write an article for the *UNSW Law Journal*, and it was here that many of the ideas in this book were first explored. We would like to acknowledge the people involved in that early article – the editors of the *UNSW Law Journal*, and the generous colleagues who provided us with feedback, including Margaret Davies and anonymous peer reviewers.

Our editor, William MacNeil, has been so supportive from the start of this book project, and we would like to thank him and the entire editorial team at Edinburgh University Press – Judith Mackenzie, Sarah Foyle, Laura Williamson, in addition to our copy editor, Eliza Wright, and indexer, Stella Reuter. We are indebted to your professionalism, attention to detail and kindness. We are also so grateful for the insightful and constructive feedback of colleagues along the (book's) way, including Nicole Graham and anonymous peer reviewers. And many thanks to our illustrator Nicholas Page, whose intricate line drawings grace the front cover and beginning of each chapter, and bring the lawful forest to visual life.

Finally, we want to acknowledge the incredible support of our families – Paul, Alexi, Charlie, Meg, David, Anne-Marie, Sarah, Myeka, Nicholas, and Henry.

Ultimately, this book is a joint project between the two of us, and its publication represents a fitting end to a rich and rare journey that began five years ago in leafy Ann Arbor. Of course, our traverse through *The Lawful Forest* has been hard work, but it has always been gratifying and deeply rewarding. Above all, it has been great fun. We will miss it.

Cristy Clark and John Page

Introduction

GINKO

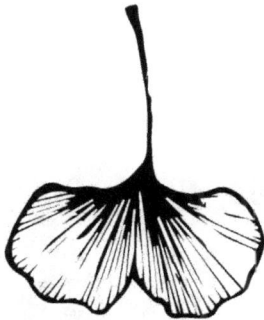

In 2016, speculative fiction writer William Gibson faced a unique dilemma. Gibson found that he was unable to finish his current book because it was set in a present that no longer existed.[1] Perversely, the future had become now. Gibson almost put the book aside for good, thinking:

> This is never going to happen. And it really threw me – for about six months. All I could do was read the news feed and feel worse about that. But eventually I realised that I wanted to believe I was living in a stub. That something had split off and that things weren't supposed to be this way. It wasn't supposed to be as dire.[2]

Gibson's sense of dread is one that many of us have shared over the last few years, a dread that has only intensified since 2016, of being blindsided by a dire turn of events and the speed with which a future dystopia seems to be approaching. Life now imitates (nightmarish) art; global pandemics, climate crises, widening social and racial inequalities, civil unrest, liberal democracy under threat, the contestation of truth, science, objective fact, these are our new 'normals'. Gibson's book was set in both the present and a post-apocalyptic future – a time after the 'Jackpot', an unspecified cataclysmic confluence of events or 'ecopolitical disasters' that wiped out 80 per cent of the human population, resulted in mass extinctions of nonhuman animals, and left the biosphere on the brink of collapse.[3] For Gibson, the evidence was in, his imagined apocalyptic *future* was his 2016 *now*.

Our book, *The Lawful Forest*, is framed for this time on the cusp, written on the precipice of Gibson's 'Jackpot'. We may be living through the

---

[1] Sam Leith, 'Interview with William Gibson: "I was losing a sense of how weird the real world was"', *The Guardian* (11 January 2020), available at <https://www.theguardian.com/books/2020/jan/11/william-gibson-i-was-losing-a-sense-of-how-weird-the-real-world-was>.

[2] Ibid.

[3] William Gibson, *Agency* (Penguin, 2020).

end-days of late-stage capitalism, when David Harvey's accumulation by dispossession nears its inexorable conclusion – and the final enclosure of our earthly commons looms.[4] All around us the portents are ominous. Species extinction accelerates, one hottest summer on record is followed by another, mega-wildfires scorch the land, and so on. Meanwhile, the human institutions we rely on for collective action seem inadequate for the herculean task. Democracy is under stress, inequalities are further entrenched, truth is relative, even insurrection surreally stalks the corridors of the U.S. Capitol. However, these dire times also speak of *other* ways forward, of past societal moments when we were on different 'edges', and of their 'shadow revolutions' rich in promise, yet never fulfilled. Our book chronicles a critical history of property, protest and spatial justice – the stories of these 'shadow revolutions' and their social movements, and the faint hopes that their prefigurative, utopian-like imaginings of an *other* future offer for these grim times.

In particular, ours is a story of space and time. It tells of our relations with space, and how one particular spatial paradigm came to dominance over time. It is *selective*, in that the stories we proffer are simply representative, not exhaustive of what we see as a linear phenomenon – one where the enclosing tendencies of the few prevail against the communitarian instincts of the many, frequently through violence and theft. Yet while linear, this trajectory is neither preordained nor inevitable. Its proponents may argue otherwise, but our book highlights what we see as points of inflection, when this excessive linearity is vulnerable to disruption, and different trajectories, other paths forward, briefly materialise before dissipating. It is also an account of the legal mechanisms that regulate and maintain this spatial linearity, the laws and theories of property, and their unequal, exclusionary and alienating effects. And in telling property's dominant narratives, this book also reminds us of property's *other* laws and customs, subordinate, relational stories that, like the physical forest, have been pushed to the periphery, overlooked yet extant. Private property, its spatial inequities and its dephysicalisation from place are, of course, not solely responsible for our current parlous precarity. We have arrived at this precipice for multiple, diverse reasons. But ours provides an account that reflects the scarce finitude of the space we share, and the enclosing, consumptive, alienating philosophies that underpin and exacerbate what has become a 'cult of exclusion'.[5] This is a 'cult' that feeds land's unequal distribution, turbocharges its unsustainable use and exploitation, and con-

---

[4] David Harvey, *The New Imperialism: Accumulation by Dispossession* (Oxford University Press, 2003) (*New Imperialism*).

[5] Nick Hayes, *The Book of Trespass: Crossing the Lines that Divide Us* (Bloomsbury Circus, 2020).

fects a dephysicalised relationship to land that is foreclosing on our future. As Nicole Graham surmises, dephysicalisation is 'one of the most significant contributions of modern Anglophonic law to anthropogenic environmental change'.[6] Or, similarly, using a Marxist analysis, 'the rupture in the material relationship between people and place could be traced directly to the expansion of private property'.[7]

So, in telling this story of space and time, our book focuses on these disruptive moments, when the linear trajectory wavered, and how the imagining of different spatial paradigms may have alleviated our current existential crisis. Its title speaks of forests that are lawful, of histories that are critical, and of the conjunction of property and protest. Each element of this title merits unpacking. First, we employ the *forest* as metaphor, a figurative expression that is both fact and signifier. Our forest is at once physical and metaphysical, a literal space of trees, woods and greenery, and a brooding metaphysical presence of forests that once were. The *metaphor* the forest represents is a certain relational understanding of people and space, an ancient spatial ordering with consequences for community, the society it emplaces and informs, and its environmental context. For most of us, the forest as fact has disappeared from mainstream view, long felled to the margins, surviving as outlier enclaves of national parks, federal forests, or remnant woods. Yet the metaphysical forest continues to linger, the imagining of an *other* way of ordering spatial relations, an aspirational, deep-seated yearning for something other than the dominant paradigm on offer. It represents, if you like, an unconsummated desire to once again make the 'forest' central to our lives. Of course, the metaphysical forest does not remain in permanent abstraction. From time to time, it takes material shape, diverse enactments of 'new ways of doing' that perform the ideal in all its utopian flaws. Community gardens, edible street verges, or food forests are contemporary examples of this yearning, green shoots that appear in the sidewalk cracks of our 'business as usual' urban landscapes. Michel Foucault employs the term 'heterotopias' to describe heterogeneous spaces, 'real places' not 'fundamentally unreal spaces'.[8] Similarly, Davina Cooper identifies 'everyday utopias', real-world sites of conceptual potency and innovation with an extraordinary potential to 'revitalize progressive and radical politics [and put] everyday concepts into

---

[6] Nicole Graham, 'Dephysicalised Property and Shadow Lands', in R Bartel and J Carter (eds) *Handbook of Space, Place and Law* (Edward Elgar, 2021) 281.

[7] Ibid 286.

[8] Cited in Edward Soja, *Postmodern Geographies: The Reassertion of Space in Critical Social Theory* (Verso, 1989) 17.

practice in counter-normative ways'.[9] Such everyday utopias and their like illustrate how the marginalised spaces and places of the metaphysical lawful forest take intermittent form, imperfect utopias that, in their flawed practices and ideals, provide a glimpse of an *other* way forward.

Second, our forest is *lawful*. While this book engages widely with history, geography, politics and other disciplinary insights, at its core it remains a scholarly legal monograph. Its tales are rooted in the common law, reflective of the institution's unremarked, overlooked plurality and the importance of social context. Hence, its lessons and narratives are *of* the law. Its lawful-ness also stems from the residue of ancient forest laws, and their related communitarian practices and customs that continue to 'haunt' the lawscape.[10] Lawful-ness also reflects these laws' collective legitimacy, derived from the 'forest floor' and validated by the ground-up actions of the many, not the prerogative top-down *diktat* of the few. In their diversity and plurality, these lawful narratives run counter to modern legal orthodoxy, where legal practice prefers a unitary, detached, politically neutral turn. Again, this messy heterogeneity renders our forest lawful, its many voices powerful in their groundswell of authenticity.

Over the last forty years of neoliberalism,[11] this modern 'orthodoxy' has been kicked into overdrive by those who those seek to profit from its dominance. Under neoliberalism's hegemonic discourse, alternative economic theories and systems of spatial ordering have been rendered implausible. This has been partly achieved by erasing the history and law that came before. One of the central tasks of this book is to disrupt this hegemony by unearthing a richer history of property's lawfulness, and all its pluralist, quirky, ancient and intriguing possibilities.

Lawful also reminds us of the need to engage with the law beyond its narrow black letter. *Legal* forests flourish and grow in their self-contained, self-referential worlds of statute, regulation and case law. This worldview makes it tempting to dismiss this shallow layer of black-letter doctrine as 'the law'. Yet the law is more 'depthy', more engrained in its stories than its black-letter thinness presupposes. In this book, we engage with critical property theory and critical history to explain and contextualise the lawful forest as a fuller picture of our legal relationships with land. Theory is especially resonant since, in the words of Margaret Davies, 'theory has an important

---

[9] Davina Cooper, *Everyday Utopias: The Conceptual Life of Promising Spaces* (Duke University Press, 2013) 13.

[10] Nicole Graham, *Lawscape: Property, Environment, Law* (Routledge, 2011).

[11] David Harvey, *A Brief History of Neoliberalism* (Oxford University Press, 2005).

role in reimagining the world and prefiguring the future'.[12] Property theory is singled out because (again channelling Davies) 'property is a particularly expansive practice of dominance [central to] present forms of global ordering and its potential to be constructively rethought'.[13] And property is, of course, central to this book's focus on space. Meanwhile, a critical perspective of history interrogates the supposed linearity of spatial 'progress' and identifies times of disruption that run counter to the straight line. These were the social movements whose varied resistances to the enclosure of space, and the social control that often accompanied it, revealed patterns of commonality. So, in later chapters, our book visits Thomas More's fictional *Utopia*, the Diggers' camps of seventeenth-century England, the Paris Commune, and the ecological communes of 1970s Australia, amongst others, to exemplify this alt-historical record. Their different yet shared resistances show how subordinate histories are marginalised. But they also underscore the potential of taking other paths, of following alternate forks in the road. In 1975, in the era of Aquarian counter-culture, emerging environmental protest, and the ongoing fallout of Paris 1968, Australia's first Minister of the Environment observed presciently that 'something is rotten at the core of conventional human existence'.[14] He encouraged a generation of young, mostly university-educated idealists to 'return to the land' and follow a 'new way forward'. Nearly fifty years on from these prophetic remarks, as the precipice nears ever closer, it is well past time to heed the lessons of these marginalised spatial realities and their counter-normative practices, theories and histories.

Finally, ours is a tale of *protest* and *spatial justice*. This again reflects the grounded truth that law is a creature of politics and power. No more so is this evident than in the laws of property. The ownership of land and its attendant rules and doctrines have long reinforced an unequal power dynamic, entrenching a spatial privilege for those of a particular race, class, or social status. As the English historian E P Thompson remarked of the eighteenth century, arguably the apogee of the enclosure period, 'land remained the index of influence, the plinth on which power was erected'.[15] For this book, land is central to this 'plinth of power', its stories exposing Jeremy Waldron's pithy

---

[12] Margaret Davies, 'Material Subjects and Vital Objects – Prefiguring Property and Rights for an Entangled World' (2016) 22(2) *Australian Journal of Human Rights* 37, 38. Davies also adds, 'at the same time, a thought of utopia – the motivation for a better if not a perfect world is intrinsic to prefigurative theory', 39.

[13] Ibid.

[14] Dr Moss Cass in M Smith and D Crossley (eds) *The Way Out: Radical Alternatives in Australia* (Lansdowne Press, 1975) 1.

[15] Hayes (n 5) 23.

truth that '[e]verything that is done has to be done somewhere. No one is free to perform an action unless there is somewhere he [sic] is free to perform it. Since we are embodied beings, we always have a location.'[16] Waldron's logic makes it plain that legal entitlements to space, whether on the rural commons of the 1700s or the urban commons of today, deeply implicate and impact on our capacity to do and to be. Property law's lines of exclusion and alienation formalise this truth. What has been lost to plain sight, however, are the ways in which space can be 'constructively rethought', where spatial access is not just the good fortune of political power or the happenstance of fate, but an expectation of spatial justice for all. Lucy Finchett-Maddock writes of 'a proprietorial right of resistance', whereby citizens engage in a 'taking back' to reclaim 'previously shared resources divided up by private property rights'.[17] This 'taking back' against the physical and metaphoric violence of enclosure is the gravamen of our interrogation of spatial *protest*, one that draws heavily on its ancient provenance. As Finchett-Maddock likewise elaborates, '[t]hese movements of resistance all desire to affect the taking of land, echoing the cries from *histories past* . . . premised on the right to land as part of an histori-cal and continual fight against dispossession.'[18] Or more eloquently, 'from the common fields of the fifteenth century, through to the abandoned spaces of the twenty-first, there is this collective mourning, a sense of loss, a ghost, forming an historical continuum from one epochal scene to the next'.[19]

In this Introduction, our tale of the lawful forest, and its critical histories of property, protest and spatial justice, begins in the American city of Detroit. The Introduction also periodically visits the uber-commoditised landscapes of Australia's cities, switching between countries as it alternates between the two extremes of private property's dominant narrative. It also lands briefly in the global city of London, where spatial inequity is never far from its unequal geographies. Detroit seems an unlikely place to launch this story, disrupting history's (comforting) linearity by starting in the *now* – then time-travelling 'backwards' in later chapters. It is also unlikely given Detroit's post-industrial setting, a context far removed from the ancient forests that permeate this book and its winnowed themes. Yet, for other compelling reasons, modern-day Detroit is as good a place to start as anywhere. In the twenty-first century, Detroit became *Detropia*, the failed city of a rich global superpower, a city

---

[16] Jeremy Waldron, 'Homelessness and the Issue of Freedom' (1992) 39 *UCLA Law Review* 295.

[17] Lucy Finchett-Maddock, *Protest, Property and the Commons: Performances of Law and Resistance* (Routledge, 2016) 121.

[18] Ibid 123 (emphasis added).

[19] Ibid 137–8.

enclosed by late-stage capitalism. It is also a city where spatial inequality and the metaphysical forest loom large. For most, the collapse of Detroit seemed a one-off, a tragedy that could be explained by its own especial, endemic factors. But with the benefit of (very recent) hindsight, perhaps Detroit was not a one-off, but a premonition writ large, a harbinger of Gibson's 'Jackpot'. Like the sight of armed Michigan Militia presaging the violent insurrections of January 2021 in the U.S. Capitol, Gibson's 'Jackpot' arrived early on the shores of Lake Michigan.

Detroit: An Allegory

Detroit's spectacular economic collapse, culminating in 2013 with the largest municipal bankruptcy filing in United States history, made the city the poster child of urban decline. Its collapse has been blamed on everything from capitalism,[20] to unions,[21] deindustrialisation,[22] racial discrimination and segregation,[23] white flight,[24] and dispersed urban planning.[25] Detroit is archetypal for all these reasons, many of which are considered in later chapters, but also because it highlights a core theme of this book: the logical consequences of enclosure and the systemic failures of private property.

Property is a constitutional bulwark of the United States, alongside life, liberty and the pursuit of happiness. Property gives its citizens 'the means to thrive . . . to exercise autonomy, to enjoy our liberties, to shape our destiny, to form relationships with others, to live a human life' as much as it provides for their material needs and private wealth.[26] Such is the republican ideal, the promised utopia. Yet in Detroit, private property did not give its citizens a means to thrive, nor to flourish. Indeed, at its nadir, it did not even provide

---

[20] Richard Wolff, 'Detroit's Decline Is a Distinctively Capitalist Failure', *The Guardian* (23 July 2013), available at <https://www.theguardian.com/commentisfree/2013/jul/23/detroit-decline-distinctively-capitalist-failure>.

[21] William K. Tabb, 'If Detroit Is Dead, Some Things Need to Be Said at the Funeral' (2015) 37(1) *Journal of Urban Affairs* 1, 3.

[22] See, e.g., Craig Freedman and Alexander Blair, 'Seeds of Destruction: The Decline and Fall of the US Car Industry' (2010) 21(1) *The Economic and Labour Relations Review* 105–26; Charles K Hyde, '"Detroit the dynamic": The Industrial History of Detroit from Cigars to Cars' (2001) 27(1) *Michigan Historical Review* 57.

[23] Heather Thompson, *Whose Detroit?: Politics, Labor, and Race in a Modern American City* (Cornell University Press, 2nd ed., 2017).

[24] Ibid.

[25] Pete Saunders, 'Nine Reasons Why Detroit Failed', *AaronRenn* (21 February 2012), available at https://www.urbanophile.com/2012/02/21/the-reasons-behind-detroits-decline-by-pete-saunders/>.

[26] Joseph Singer, *The Edges of the Field: Lessons on the Obligations of Life* (Beacon Press, 2000) 27.

for their material security. Rather, property delivered a distinctly private tragedy. Detroit's path to collapse is well documented and the narrative familiar. Once the dominant manufacturing powerhouse in the United States – from the second half of the nineteenth century through to the 1970s – Detroit fell into an increasingly rapid decline when its cornerstone automobile industry imploded.[27] As author Drew Philp neatly observes of its post-industrial ruin, '[t]he city that put the world on wheels drove away in the cars they no longer made.'[28]

Those driving away were mostly white Detroiters. Detroit's first wave of deindustrialisation, coupled with highly discriminatory housing policy and practice, trapped its Black and minority citizens in an increasingly empty, poverty-stricken city. Between 1970 and 2010, the city lost over half its population, going from a high of 1,514,063 to 713,777.[29] By 1990, Detroit was ranked 'first' in poverty, 'had the highest percentage of households receiving public assistance payments' and was at 'the bottom in terms of the median value of owner-occupied homes'.[30] In its manufacturing boom years, Detroit was also home to the strongest labour movement in the United States, with the United Auto Workers (UAW) playing a leading role in 'lobbying for [and securing] progressive social and economic legislation'.[31] Conservative accounts of Detroit's collapse blame the UAW for making domestic cars uncompetitive in a globalised market, while liberals counter that 'by the 1960s the multi-story, huge plants in the city had become obsolete, replaced by far larger single-story plants built outside the city, often in areas with lower wages and non-union local labor . . .'.[32] Over time, neoliberal policies of enforced 'urban entrepreneurialism',[33] coupled with significant reductions in state revenue sharing, reversed progressive laws and enlivened conservative narratives, and ultimately led to complete economic collapse.[34] Harvey describes this entrepreneurial governance model, with its promotion of public–private partnerships and their shifting of speculative private risk to the public sector, as exerting an 'external coercive power over

[27] See, e.g., Freedman and Blair (n 22) 105; Hyde (n 22) 57.
[28] Drew Philp, *A $500 House in Detroit: Rebuilding an Abandoned Home and an American City* (Scribner, 2017) 3.
[29] Thompson (n 23) xv.
[30] Tabb (n 21) 1.
[31] Ibid 2.
[32] Ibid.
[33] David Harvey, 'From Managerialism to Entrepreneurialism: The Transformation in Urban Governance in Late Capitalism' (1989) 71(1) *Geografiska Annaler B* 3, 12 ('From Managerialism to Entrepreneurialism').
[34] M H Desan, 'Bankrupted Detroit' (2014) 12(1) *Thesis Eleven* 122.

individual cities to bring them closer into line with the discipline and logic of capitalist development'.[35] Another key feature of this financialisation is the 'drift towards post-democratic technocratic management', best exemplified by the imposition of unelected Emergency Managers who cannot be held accountable by residents.[36]

The gathering drift to collapse did not go unchallenged. Waves of protest, spontaneous, grass roots, and often led by the Black majority also form a significant part of Detroit's history. The labour movement played a significant early role in these protests – fighting for workers' rights, equality and class solidarity, and helping to establish a 'New Deal'.[37] But, in an echo of an ongoing tension between bounded and unbounded claims to the commons that this book explores, these solidarity gains were undermined by persistent racial inequality and the reluctance of white union members to give up 'the prerogatives of their whiteness'.[38] This reluctance took particular form in the activism of white homeowners' associations in fighting the incursion of Black neighbours and, ultimately, in the white flight that served to hollow out the city. Such measures disproportionately favoured white Detroiters,[39] sparking a new round of protests, including the infamous race riots of July 1967. In this dramatic five-day protest, black Detroiters rebelled against the structural racism that systematically excluded them from the promised prosperity of private property and the *American Dream*.[40] Protest and Finchett-Maddock's 'proprietorial right of resistance'[41] are deeply engrained in Detroit's DNA.

Detroit's allegory of collapse is a potent symbol of *dystopia*. Dystopia was the genesis of the 2012 film, *Detropia*, 'a cinematic tapestry featuring the lives of several citizens trying to survive the D and make sense of what is happening to their city'. In *Detropia*, the 'birthplace of the [American] middle class' is said to face a dystopic future, where 'racial tension, globalization, lack of innovation, and corporate greed has led to a moment of truth'.[42] Utopia and dystopia are two ends of a spectrum canvassed in this book, the two extremes of property's role in place making, the sublime and the surreal. Detroit's

---

[35] Harvey, 'From Managerialism to Entrepreneurialism' (n 33) 10–11.

[36] Jamie Peck and Heather Whiteside, 'Financializing Detroit' (2015) 92(3) *Economic Geography* 235, 238.

[37] Thomas J Sugrue, 'Labor, Liberalism, and Racial Politics in 1950s Detroit' (Fall 1997) 1 *New Labor Forum* 19; Thompson (n 23).

[38] Sugrue (n 37) 23.

[39] Ibid.

[40] Thompson (n 23) ix.

[41] Finchett-Maddock (n 17) 130.

[42] Heidi Ewing and Rachel Grady, *Detropia* (2012), available at <https://detropiathefilm .com>.

dystopia yielded its own bespoke *Detropia*, in what may have been an early manifestation of the 'Jackpot'.

Central to dystopia is the multilayered concept of *alienation* (canvassed in greater detail in Chapter 1). In its lay sense, alienation is exemplified by the social, economic and environmental isolation that Detroit's citizens felt from their home city – as its economy collapsed and communities were hollowed out by dramatic population loss. Likewise, the city itself was *alienated*, detached from its natural landscape through a toxic legacy of decades of unchecked industrialisation that polluted its land and waters.[43] And in its car-centric planning, Detroit was the classic 'sprawlscape', the dispersed metropolis that exploded across the countryside in the post-war years, a city designed for capital and the almighty automobile. Much of Detroit fits the image of James Howard Kunstler's 'geography of nowhere',[44] the anonymous, *alienating* landscapes dominated by dispersed private housing, mega-shopping malls, few sidewalks and fewer parks. In these geographies of nowhere – places that seem like *anywhere* in their placeless-ness – the public square is largely absent or impoverished. Apart from the ubiquitous freeway – which connects one far-flung suburb to another, there is a dearth of public infrastructure amidst a private plenitude. As urban planner Pete Saunders describes his hometown:

> Detroit's streetscape is unbearable in many places. Major corridors have long stretches of anonymous single-story commercial buildings, with few trees or other landscaping. Signs, banners, awnings and decorative lighting are noticeably lacking. Overhead electrical wires extend for miles, and streets have been rigidly engineered with road signs and markings.[45]

When faced with this wasteland of alienation, the policy-makers' answer to Detroit's collapse was more of the same. This autopilot response embodies another understanding of alienation: the right of private capital to alienate – to commodify – and in Detroit's case, a prescription to alienate its way out of malaise. This interpretation of alienation is well understood by property

---

[43] Saunders (n 25); Terressa A Benz, 'Toxic Cities: Neoliberalism and Environmental Racism in Flint and Detroit Michigan' (2017) *Critical Sociology*; Liam Downey, 'Environmental Racial Inequality in Detroit' (2006) 85(2) *Social Forces* 771; Cristy Clark, 'Race, Austerity and Water Justice in the US: Fighting for the Human Right to Water in Detroit and Flint, Michigan', in F Sultana and A Loftus (eds) *Governance, Rights, and Justice in Water: New Ideas and Realities* (Routledge, 2019).

[44] James Howard Kunstler, *The Geography of Nowhere: The Rise and Decline of America's Man-Made Landscape* (Simon & Schuster, 1994).

[45] Saunders (n 25).

scholars, a core right to sell, dispose, mortgage, and in countless other ways, deal with the private 'bundle of sticks' that Detroit had become. Alienation's centrality draws our attention to the role that private property and enclosure played in Detroit's demise. While the causes of Detroit's collapse were multiple and complex, as preceding paragraphs infer, it remains true that the extreme alienation that played out in the city is a logical endpoint of private property. Indeed, Detroit could be (yet again) the uber poster child of Eduardo Peñalver's theory of 'property as exit', the liberal paradigm that emphasises the paramountcy of the individual and views community with suspicion, an entirely optional intermediary between the welfare-maximising individual and the minimalist regulatory state.[46] Where property is exit from place, values of alienation and commodification overwhelm property's relational 'other-half', permitting wealthier residents to vote with their feet (or drive away in their cars), trading up without regard to consequences for communities left behind. The role private capital played in Detroit in facilitating exit can be traced through numerous state interventions or omissions, including state subsidies to private manufacturing, agreements that reduced labour power, government programmes to encourage white suburban home ownership, dispersed urban planning, predatory mortgage lenders and subsequent foreclosures, and the enforced shift to urban entrepreneurialism.[47] Detroit's modern-day tragedy was no tragedy of the commons (the polemic thesis expounded by Garrett Hardin in 1968, critiqued in Chapter 1), but its antithesis, a very private tragedy of unchecked alienation, one that played out on common ground and in full public view. In Detroit, capitalism and private property metastasised in what was (and is) a self-consuming tragedy of the logic of enclosure.

Yet amidst the devastation of this hyper-enclosure and exit in Detroit, vacant space was (literally) created for an *other* way of thinking about property and place. The loss of around 1.3 million residents left empty, hollowed-out neighbourhoods, especially on the city's outer edges, where so-called 'urban prairies' characterised by empty streets and 'old defunct street lamps stretched across waist-high brush' now proliferate.[48] Some of these abandoned or derelict parcels have been reclaimed and repurposed – as urban farms, street orchards, guerrilla gardens and community woods. Indeed, urban agriculture is but one manifestation of what Claire Herbert terms the rise of 'property informality' in the declining city. 'Squatting, blotting ("squatting the block"),

---

[46] Eduardo M Peñalver, 'Property as Entrance' (2005) 91(8) *Virginia Law Review* 1889.
[47] Peck and Whiteside (n 36); Tabb (n 21).
[48] Claire W Herbert, *Urban Decline and the Rise of Property Informality: A Detroit Story* (University of California Press, 2021) 35.

demolition, scrapping, salvaging and art projects' are likewise commonplace.[49] These diverse spatial tenures act as the converse of exit from place, reminding us that property is also an entrance to community,[50] locales where context and acts of relational performance forge bonds between people and place. In Detroit, these relational performances of property are explored though the lens of a private urban forest project and two urban farms. As these examples illustrate, this phenomenon is complex and multilayered. At one level, it suggests 'property as commodity' is not uniquely unidirectional. Community pushback against enclosure, this instinctive resistance to private overreach, is an equally ancient and ongoing story. However, these examples also strike a cautionary note. They illustrate how the rhetoric around community claims to space can be manipulated or compromised.[51]

The Hubris of Private Property

As flagged, this book explores our fascination with the forest, and its potent symbolism. In the post-industrial 'wastelands' of Detroit, the forest seems far away, alienated from the city and alienating in its absence. Yet for us as first-time travellers flying into Detroit's Metropolitan Airport, the city's urban blight was lost to plain sight, an urbanity surprisingly softened by the early spring foliage of its forests of deciduous trees.

Once we had landed, it became clear that this 'forest' cover hides a multitude of sins. Yet Detroit's malaise presented (and continues to present) much private opportunity. Waterfront redevelopments, new sport stadiums,[52] ever-bigger shopping malls, public–private partnerships, even a proposed zombie theme park,[53] offered private panaceas for a distinctly private disease. In the so-called 7.2, referring to an area of 7.2 square miles comprising the precincts

---

[49] Ibid 2.

[50] Peñalver (n 46).

[51] For example, Herbert observes how the provenance of urban agriculture is contested, with newcomer, mostly white residents (so-called 'Lifestyle Appropriators', 'thirty-somethings with University of Michigan degrees and comfortable suburban upbringings') using the language of 'homesteading' and other settler constructs to describe their 'pioneering' urban farms, while Black residents reject the idea of its 'reintroduction', citing long-established urban farming practices that date back to the 'southern roots of African Americans and other groups', Herbert (n 48) 190, 205.

[52] For example, the hockey stadium built on the city's waterfront for Red Wings owner, Mike Illich, a 'Detroit billionaire pizza magnate'. There the City contributed 60 per cent of the construction costs, P E Moskowitz, *How to Kill a City: Gentrification, Inequality and the Fight for the Neighborhood* (Bold Type Books, 2018) 97.

[53] Seth Schindler, 'Detroit after Bankruptcy: A Case of Degrowth Machine Politics' (2016) 53(4) *Urban Studies* 818, 825.

of Midtown and Downtown, the city is booming, a publicly subsidised zone of gentrification that 'skipped the beginning phase with the artsy folks, the laid-back coffee shops, and the activists, and instead jumped straight from broke dystopian metropolis to yuppified playground'.[54] Meanwhile, the remaining 134.8 square miles of the city 'slowly falls off the map, bled out by foreclosures, blight, and a lack of city services'.[55] Walking through this blighted landscape and witnessing the empty, boarded-up houses – many of which had clearly been quite grand in their day – was a shocking experience for us as Australians, where the hyper-commodified housing market at home leaves no material or conceptual space for such urban wastelands.[56] This stark contrast belies a deeper truth, that these polar extremes are in fact outcomes of the same ideology and process.

Such is capitalism's hubris that the policies of privatisation and enclosure responsible for creating and exacerbating Detroit's financial woes were repackaged as key elements of the proposed path out of bankruptcy, or at least, for the mostly white elite part that covers around 3 per cent of the city's footprint. Indeed, perhaps at its most ironic was one developer's proposal for a private urban forest. In this example, we find the perfect convergence of this book's central motifs, where the private estate's insatiable urge to enclose public resources intersected with the very idea and materiality of the forest.

In 2008, the investor John Hantz approached the city government with a proposal to revitalise a neighbourhood in Detroit's East Side by turning over 1,500 lots into an urban forest, the trees to be planted with the aid of thousands of community volunteers.[57] His company, the Hantz Group, argued that the project would benefit everyone – providing cash for the city, clearing derelict properties and trash from vacant blocks, making the neighbourhood more liveable, and acting as a profitable investment for the group.[58]

---

[54] Moskowitz (n 52) 74. The 7.2 is described as 'a dome of luxury tightly sealed at its borders', ibid 120.

[55] Ibid 91. Real estate developers in Detroit use the term 'blank slate' to describe the city, a *terra nullius* that 'ignores not only the 700,000 other people who still live there, but also the historical reality that the "blank slate" was created through brutal racism', ibid 116.

[56] As Herbert similarly observes, property functions 'very differently under conditions of growth versus decline. Under the former, property is in high demand, low supply and often increasing in value . . . But under conditions of decline, property is in abundant supply, holds little economic value, and is often a liability more than an investment.' It is in these latter 'interstitial, poorly regulated spaces' that property informality flourishes, Herbert (n 48) 4, 5.

[57] See, e.g., Flaminia Paddeu, 'Legalising Urban Agriculture in Detroit: A Contested Way of Planning for Decline' (2017) 88(1) *The Town Planning Review* 109.

[58] Sarah Goodyear, 'A 140-Acre Forest Is about to Materialize in the Middle of Detroit',

The project was controversial, with local residents arguing that it was a 'land grab'.[59] Sara Safransky has described it as 'accumulation by green dispossession' – a process by which collective land and resources are appropriated on the basis of 'the expendability of particular people and places'. Safransky argues that the Woodlands Project is part of a larger pattern of dispossession in Detroit, devised by the City under the fifty-year Detroit Future City (DFC) plan. Under this plan, Detroit's so-called 'high-vacancy neighbourhoods' were deemed surplus to requirements, and city planners proposed to deal with urban blight and Detroit's overly dispersed spatial footprint by withdrawing urban services and effectively rendering certain areas uninhabitable.[60] As these areas were 'deemed to have no market value', they were slated for blue-green infrastructure, including Hantz's private urban forest.[61]

Despite this early controversy amongst residents,[62] media reception of the Woodlands Project was largely uncritical.[63] Its appeal seemed to lie in two key promises: that it would clean up abandoned land at no cost to the city, and that it would create a *forest*. There is a reason the word 'forest' was used in the promotion of the project. A central argument of this book is that 'forest' is code for a relational understanding of property. It carries a deep symbolic meaning, because it echoes an early memory of an *other* way of relating to property and place.

CityLab (25 October 2013), available at <https://www.citylab.com/design/2013/10/140-ac re-forest-about-materialize-middle-detroit/7371/>.

[59] See, e.g., Morgan Linn, '"Land grab" Documentary Follows Story of Controversial Blight Removal in Detroit', *Great Lakes Echo* (1 March 2017), available at <https://greatlakesecho .org/2017/03/01/land-grab-documentary-follows-story-of-controversial-blight-removal-in -detroit/>.

[60] Sara Safransky, 'Greening the Urban Frontier: Race, Property, and Resettlement in Detroit' (2014) *Geoforum* 237–48.

[61] Ibid 237. 'Right-sizing' is another (earlier) descriptor for this process, the deliberate shrinking back to a spatial scale where 'the burdens of infrastructure costs, blight, and vacancy would cease, and the city would once again be able to take care of itself' through the permanent withdrawal of city services, Herbert (n 48) 19.

[62] See, e.g., Paddeu (n 57); David Sands, 'Hantz Farms Deal, Controversial Land Sale, to Go before Detroit City Council', *Huffington Post* (20 November 2012), available at <https:// www.huffingtonpost.com.au/entry/hantz-farms-deal-land-detroit-council_n_2159863>.

[63] See, e.g., John Gallagher, 'Why Is a Person Cleaning Up Detroit Getting the Runaround from the City?', *Detroit Free Press* (19 June 2016), available at <https://www.freep.com/sto ry/money/business/columnists/2016/06/19/hantz-detroit-woodlands-greening-duggan/85 979486/>; Goodyear (n 58); Bill Bradley, 'Debunking the Myth of Detroit's New Urban Forest', *Next City* (31 October 2013), available at <https://nextcity.org/daily/entry/debun king-the-myth-of-detroits-new-urban-forest>.

In the end, no forest materialised – the trees are quite dispersed. But, by 2018, with the purchase and clean-up of some 1,500 parcels of 'blighted land' and the planting of some 25,000 trees, the Woodland project was largely complete. The neighbourhood has reportedly become more liveable as a result, with safety and amenity improving.[64] Land value has also increased. Nonetheless, the Hantz Woodlands Project has failed to function as a genuine solution to Detroit's woes. More of the same, even when dressed up as a forest, does nothing to solve the fundamental issues of alienation and commodification. The core problems with such blind adherence to market-based solutions are thrown into stark relief by the example of Australia's hyper-commodified housing market, where the seemingly exponential growth in 'values' has locked out a generation, alienated lower-income communities from the inner suburbs of Sydney, Melbourne and other cities, large and small, and led to a crisis of homelessness and housing stress.[65]

This issue of hyper-commodification came to the fore in 2017 when the New South Wales state government passed legislation empowering police to dismantle a protest camp in Martin Place – a pedestrian mall in Sydney's CBD.[66] The camp's location, directly outside the Reserve Bank in the heart of the city's financial and legal district, was deliberate. It was established by advocates to serve two related purposes: the first to provide a form of 'housing' for the growing homeless population of Sydney; the second to act as a material protest against the policies that lead to homelessness. The rapid dismantling of the camp and its communal facilities – its effective erasure – was deeply symbolic. At one level, it exemplifies Nicholas Blomley's argument that private property must relentlessly (and ruthlessly) prosecute its core message. This is a battle with no end in sight, daily struggles for propertied dominance that must be won lest *other* propertied discourses gain the slightest foothold.[67] At another level, this act of erasure also reflects the discomfort

[64] See, e.g., Matthew Smith, 'Hantz Farms Seeing New Success with Urban Farming in Detroit', *WXYZ Detroit* (30 June 2017), available at <https://www.wxyz.com/news/hantz-farms-seeing-new-success-with-urban-farming-in-detroit>.

[65] Jessie Hohmann, 'Toward a Right to Housing for Australia: Reframing Affordability Debates through Article 11(1) of the International Covenant on Economic, Social and Cultural Rights' (2020) 26(2) *Australian Journal of Human Rights* 292–307, DOI: 10.1080/1323238X.2020.1813378.

[66] Cristy Clark, 'Clearing Homeless Camps Compounds the Violation of Human Rights and Entrenches the Problem', *The Conversation* (11 August 2017), available at <https://theconversation.com/clearing-homeless-camps-compounds-the-violation-of-human-rights-and-entrenches-the-problem-82253>.

[67] Nicholas Blomley, *Law, Space, and the Geographies of Power* (Guilford Press, 1994) (*Law, Space, and Power*).

of property owners with the provenance of their title. Homelessness is an uncomfortable reminder of property's frailties and flaws. Carol Rose describes this discomfort as an 'ownership anxiety',[68] especially potent in the colonial context where historic (and ongoing) land thefts are grounded in a fiction of *terra nullius*. Such a commodified approach to land is also at odds with Indigenous worldviews, where the law emanating from Country carries deep relational obligations.

Meanwhile, back in Detroit, Hantz was always open about his intention to profit from the project. In 2010, he told *The Atlantic*:

> We need to create scarcity, because until we get a stabilized market, there's no reason for entrepreneurs or other people to start buying. I thought, What could do that in a positive way? What's a development that people would want to be associated with? And that's when I came up with a farm.[69]

In May 2018, brochures appeared at an Urban Land Institute conference for 4,000 real estate developers, brokers and investors advertising development opportunities within the 188 acres owned by the Hantz Group.[70] When approached about whether they were selling off the land, Mike Score, president of Hantz Woodlands LLC, claimed, 'It's not like we are heading off in a new direction. We are trying to get information about what impact our investment has had on neighborhoods within Hantz Woodlands. We are measuring marketplace response to the investment we made.'[71]

While liveability may well have improved for locals, for now, control over their neighbourhood has been placed in the hands of one, unelected man who has been consistently open about his entrepreneurial goals. The Hantz Group's profits also come off the back of volunteer labour and community energy. There is something telling about the fact that all this has been enclosed into private property. While Hantz retains perfect security of tenure over his Woodlands, the community's rights are fragile.

The underlying narrative of the Hantz Woodlands Project, and other government projects to revitalise Detroit, is that the way out of the collapse is more of the same. Even as the system unravels, its logic of enclosure remains

---

[68] Carol Rose, 'Canons of Property Talk, or, Blackstone's Anxiety' (1998) 108 *Yale Law Journal* 601.

[69] Eleanor Smith, 'John Hantz', *The Atlantic* (November 2010), available at <https://www.th eatlantic.com/magazine/archive/2010/11/john-hantz/308277/>.

[70] Kirk Pinho, 'Is Hantz Farms Property Up for Sale? Owner Says No', *Crain's Detroit Business* (6 May 2018), available at <http://www.crainsdetroit.com/article/20180506/news/659981 /is-hantz-farms-property-up-for-sale-owner-says-no>.

[71] Cited in ibid.

unquestioned. Many scholars have argued that capitalism's resilience in the face of inevitable crises is partly due to the seemingly paradoxical accepted logic that the solution to these crises is to intensify the very policies that caused them in the first place.[72] Naomi Klein has called this 'the Shock Doctrine',[73] a prescriptive approach to disaster capitalism that was first practised in the global south before being imported to cities in the global north. Jamie Peck describes this repackaged approach as 'austerity urbanism'.[74] With austerity urbanism,

> [pressures] operate downwards in both social and scalar terms: they offload social and environmental externalities on cities and communities, while at the same time enforcing unflinching fiscal restraint by way of extralocal disciplines; they further incapacitate the state and the public sphere through the outsourcing, marketization and privatization of governmental services and social supports; and they concentrate both costs and burdens on those at the bottom of the social hierarchy, compounding economic marginaliza-tion with state abandonment.[75]

Performing Belonging and the *Other* in Contested Landscapes

Across town from the Hantz Woodlands is the Michigan Urban Farming Initiative (MUFI) – a registered not-for-profit that runs an urban agriculture project based on vacant land nestled within Detroit's North End. Like the Hantz forest farm, MUFI uses urban agriculture, along with other com-munity projects, to improve the amenity and value of its surrounding neigh-bourhood. MUFI was co-founded in 2011 by two University of Michigan students, Tyson Gersh and Darin McLeskey, from the nearby college town of Ann Arbor. In contrast to Hantz, Gersh, who now runs MUFI alone as President, does not own the underlying property – instead, part of the site is owned by MUFI and part is still owned by the City.

MUFI got its start under an adopt-a-lot programme, but as it expanded it encountered resistance.[76] By 2016, MUFI was still unable to obtain the deed for one of its properties or buy several others from the Detroit Land

---

72 Schindler (n 53) 818.
73 Naomi Klein, *The Shock Doctrine: The Rise of Disaster Capitalism* (Penguin, 2007).
74 Jamie Peck, 'Austerity Urbanism: American Cities under Extreme Economy' (2012) 16(6) *City* 625.
75 Ibid 650–1.
76 Christine Ferretti, 'Urban Farms Flourish, but Neighbors Feel Growing Pains', *The Detroit News* (13 October 2016), available at <https://www.detroitnews.com/story/news/local/det roit-city/2016/10/13/detroit-urban-farming/92031400/>.

Bank Authority.[77] While Gersh expressed frustration at these roadblocks to MUFI's expansion, this was not a case of needless bureaucracy.[78] The fact is that MUFI's inner-north location sits close to the last stop of the QLine light rail system and Wayne State University, and the City has sought to prioritise the development of affordable housing within this zone.[79] As Vince Keenan, neighbourhood manager for District 5 (where MUFI is located), explained:

> It's one of the few places we would be able to target affordable housing that would have easy access to the train and that's our priority. We feel it's our obligation to try to make sure people of all income levels have an opportunity to live near an amenity like that.[80]

During this time, MUFI had about a dozen projects running on its 1.5-acre site and had supplied more than 50,000 pounds of free produce to neighbourhood residents, area churches and food pantries.[81] Then, in November 2016, it announced plans to expand the site into an urban 'agrihood' by using the existing farm 'as the centerpiece of a mixed-use development'.[82] Gersh argues (using the rhetoric of commodification) that this project will help MUFI to continue to attract 'new residents and area investment'.[83] Although seemingly cast from a different mould to Hantz Farms, this focus on development and now even international tourism and real estate values seems remarkably similar and has led some to accuse Gersh of neocolonialism.[84]

The most problematic aspect of Gersh's focus on property values is the risk that gentrification poses to long-term residents, many of whom fear being forced out due to rising rents. When coupled with the fact that Gersh is a young white man from a nearby wealthy college town, the result has been competing narratives as to the value of MUFI to existing residents and

---

[77] Ibid.

[78] Ibid.

[79] Ibid.

[80] Vince Keenan, cited in ibid.

[81] Chris Ehrmann, 'Michigan Urban Farming Initiative Grows Plan for "Agrihood" in Detroit', *Crain's Detroit Business* (1 December 2016), available at <https://www.crainsdetro it.com/gallery/nonprofit/michigan-urban-farming-initiative-grows-plan-for-agrihood-in -detroit>; Ferretti (n 76).

[82] The Michigan Urban Farming Initiative, 'America's First Sustainable Urban Agrihood Debuts in Detroit', The Michigan Urban Farming Initiative Press Release (30 November 2016), available at <http://www.miufi.org/america-s-first-urban-agrihood>.

[83] Ibid.

[84] Tom Perkins, 'On Urban Farming and "Colonialism" in Detroit's North End Neighborhood', *Detroit Metro Times* (20 December 2017), available at <https://www.met rotimes.com/detroit/on-urban-farming-and-colonialism-in-detroits-north-end-neighborho od/Content?oid=7950059> ('On Urban Farming and "Colonialism"').

competing perceptions of the community that it claims to serve.[85] These complexities are nothing new. The global and historical record explored in this book reveals countless tales of enclosure and resistance, as well as contested notions of belonging and community.

An example of these contested notions of belonging and community can be seen in the case filed by a Black Detroit man, Marc Peeples, against three white women for repeatedly fabricating police reports against him.[86] Peeples had been building an urban garden in Hunt Park in Detroit's east, when Deborah Nash, Martha Callahan and Jennifer Morris – who reportedly had their own plans for the green space – sought to have him driven out by calling the police and alleging that he was engaged in criminal conduct.[87] In March 2018, for example, Nash called the police claiming that Peeples was carrying a gun. Police arrived to find him raking leaves in his Hunt Park garden instead. After this, the women's accusations escalated to include claims of violent threats and even paedophilia.[88] Although police were recorded on camera dismissing the allegations as 'BS', Peeples was ultimately arrested only to be acquitted by a Wayne County Circuit judge, who found the women's allegations were fabricated and largely motivated by race.[89] Peeples sued, seeking damages from the women whose actions led to him being arrested, locked up and reputationally damaged – all for the crime of 'gardening while Black'. He has since returned to his garden, now called Liberated Farms, and settled the litigation.[90]

This incident foreshadowed a 2020 incident in a section of New York City's Central Park known as 'the Ramble', where Christian Cooper was birdwatching. When he asked Amy Cooper (no relation) to put her dog on a leash, she responded by calling the police and fabricating claims that 'an African American man [was] threatening [her] life'.[91] The incident was

---

[85] Ibid.

[86] Tom Perkins, 'Detroit Man Sues Three White Women Who Called Police on Him for "gardening while Black"', *Detroit Metro Times* (4 March 2019), available at <http://metro times.com/table-and-bar/archives/2019/03/04/detroit-man-sues-three-white-women-who -called-police-on-him-for-gardening-while-black>; Tanasia Keeney, 'Detroit Man Files Suit against White Women Who Had Him Arrested for "gardening while Black" and other Bogus Allegations', *Atlanta Black Star* (5 March 2019), available at <https://atlantablackst ar.com/2019/03/05/detroit-man-files-suit-against-white-women-who-had-him-arrested -for-gardening-while-black-and-other-bogus-allegations/>.

[87] Ibid.

[88] Ibid.

[89] Ibid.

[90] Ibid.

[91] Jasmine Aguilera, 'White Woman Who Called Police on a Black Man at Central Park

recorded on camera and took place on the very same day as the murder of George Floyd by Derek Chauvin, a Minneapolis police officer, which was also captured on film. Both videos 'went viral' and sparked a resurgence of Black Lives Matter (BLM) protests across the United States (and elsewhere). Both the incidents and BLM highlight the widespread denial of the right of so many to safely exist in public space – particularly Black Americans.

This issue of access to public space came under increased scrutiny during the coronavirus pandemic when lockdowns and related restrictions on freedom of movement highlighted the highly uneven distribution of parks and green spaces for urban residents. In London, George Monbiot observed that spatial inequities had become unambiguously clearer, demonstrably starker. At the height of early lockdowns, he described the plight of Londoners as 'swelter[ing] in tiny flats, or edg[ing] round each other in miniscule parks, desperate for a sense of space and freedom'.[92] In England, where 92 per cent of land remains off-limits, this experience is 'nothing new'. Public health lockdowns shine a bright light on an old normal – that 'we've been kept off the lands for centuries'.[93] Monbiot's remedy encompasses an enlarged urban right to roam; where legislated access to the 11,000 acres of Greater London's 131 private golf courses, closed school grounds, or locked private green squares would dramatically expand the metropolis's common-wealth. Such imagery of 'miniscule parks' around which urbanites anxiously 'edge' depicts an ancient narrative of spatial inequality. But in the densely crowded cities of an airborne pandemic, the lawful forest is not only a visceral reminder of spatial injustice; it is also an urgent cry for healthy space as a public right, not a private privilege.

In the inner-north suburb of Northcote in Melbourne, Australia, residents responded to this issue of spatial injustice with direct action by cutting through the perimeter fence of the local golf course to gain community access during lockdown. To shore up their claims of legitimate use, the community organised picnics and other performative events in ways that echo legal geographer Nicholas Blomley's thesis in 'Performing Property, Making the World'.[94] This simple act of taking back public space for public benefit led to

---

Apologizes, Says "I'm not a racist"', *Time* (26 May 2020), available at <https://time.com/58 42442/amy-cooper-dog-central-park/>.

[92] George Monbiot, 'Lockdown Is Nothing New: We've Been Kept Off the Land for Centuries', *The Guardian* (22 April 2020), available at <https://www.theguardian.com/com mentisfree/2020/apr/22/lockdown-coronavirus-crisis-right-to-roam>.

[93] Ibid.

[94] Nicholas Blomley, 'Performing Property, Making the World' (2013) 27 *Canadian Journal of Law and Jurisprudence* 1.

heated debates around whether this newly designated 'people's park' should be repurposed, and whether or not golfers should share the space with other uses and public purposes, including rewilding.[95]

The BLM movement also emphasises that these wider issues of spatial justice are further exacerbated by the ongoing legacy of structural racism, unequal access to public space, and the ways that it continues to shape notions of belonging. MUFI's charity-based approach, embodied by its commitment to handing out free food, has been criticised for failing to address this ongoing legacy and the underlying disparities of power that underpin Detroit's social malaise.[96] As Devita Davison, Executive Director of FoodLab Detroit, made explicit in her 2017 critique, 'I'm exasperated by disillusioned white-led orgs like MUFI that profess that simply creating access to food is the answer to alleviating chronic hunger and poverty, when [in] actuality, they're sustaining the structural inequities of power that characterize it.'[97]

Held up in contrast to MUFI are urban agriculture projects like Oakland Avenue, Feedom Freedom and D-Town Farm, which have been created and driven by local community members with a focus on self-determination, social justice and racial equality. D-Town Farm, for example, is a 7-acre urban farm run by the Detroit Black Community Food Security Network (DBCFSN). Founded in 2006 by Black liberation activist Malik Yakini, DBCFSN's mission is to work for community-led food security, education, collective action and community building.[98] Given his background in Black activism, Yakini was also motivated by a desire to see Detroit's majority Black community take the lead in urban agriculture and renewal projects in the city.[99]

DBCFSN leases the site of the D-Town Farm from the City for one dollar a year, but two earlier production sites had to be discontinued when

---

[95] Melissa Davey, 'Fair Way? Covid Turned a Melbourne Golf Course into a Public Park and Now No One Wants to Leave', *The Guardian* (17 October 2020), available at <https://www.theguardian.com/australia-news/2020/oct/17/fair-way-covid-turned-a-melbourne-golf-course-into-a-public-park-and-now-no-one-wants-to-leave>.

[96] Perkins, 'On Urban Farming and "Colonialism"' (n 84).

[97] @DevitaDavison (Twitter, 20 December 2017, 1:52am), available at <https://twitter.com/DevitaDavison/status/943131843330822145>.

[98] Monica M White, 'D-Town Farm: African American Resistance to Food Insecurity and the Transformation of Detroit' (2011) 13(4) *Environmental Practice* 406; Jan Richtr and Matthew Potteiger, 'Farming as a Tool of Urban Rebirth? Urban Agriculture in Detroit 2015: A Case Study', in G Cinà and E Dansero (eds) *Localizing Urban Food Strategies. Farming Cities and Performing Rurality*, 7th International Aesop Sustainable Food Planning Conference Proceedings, Turin, 7–9 October 2015 (Politecnico di Torino, 2015) 463, 468–9.

[99] Richtr and Potteiger (n 98) 463, 468.

the first was purchased by a developer and the second was reclaimed for use by the church that owned it.[100] The farm grows over thirty different fruits, vegetables and herbs and sells them locally at farmers markets and to wholesale customers to fulfil its mission of providing healthy fresh food to residents of Detroit, which has long been classified as a 'food desert'. Unlike MUFI, which relies exclusively on volunteer labour, DBCFSN provides paid employment on D-Town Farm.[101] It has also taken the lead in advocating for food security policy and law reform, in addition to engaging in a wide range of other community development activities, including a food co-operative.[102]

Both MUFI and D-Town receive financial support from competitive grants and corporate sponsorship. Here, they are leveraging the appeal of their *otherness*,[103] their relational approach to property and place to attract support for a community in decline. MUFI and D-Town both represent attempts to create Davina Cooper's *everyday utopias*, 'material practices and spaces . . . conceptually potent, innovative sites that can revitalize progressive and radical politics through their ability to put everyday concepts into practice in counter-normative ways'.[104] Everyday utopias unsettle orthodoxies because they seek to 'understand property not through private ownership, but through [utopian perspectives of] squatting, common or public lands'.[105] Their early informal regulation also reminds us of the ever-present reality of legal pluralism – that multiple legal orders can exist simultaneously and can be created from the ground up.

Urban agriculture, protest camps, reclaimed golf courses and Liberated Farms all typify how diverse property claims can be crafted from context. Blomley argues that such performances are acts of defiance against private hegemony. Property expressed in this relational and performative way represents a form of social tenure and 'a battle for resources, a battle against scarcity of land, environmental abuse and wasted opportunities, . . . a fight for freedom of expression and for community cohesion'.[106] These are the lessons of our selective tour through the contested landscapes of Detroit, Sydney, London and Melbourne.

[100] Detroit Black Community Food Security Network, 'About Us' (2019), available at <https://www.dbcfsn.org/about-us>.
[101] Ibid.
[102] Ibid; Detroit People's Food Co-Op, 'Mission & Purpose' (2021), available at <https://detroitpeoplesfoodcoop.com/about-us/>.
[103] Andre van der Walt, 'The Marginality of Property', in G Alexander and E Peñalver (eds) *Property and Community* (Oxford University Press, 2010).
[104] Cooper (n 9) 11.
[105] Ibid 32.
[106] Ibid 181–2.

Disrupting Property's Orthodoxy

It is the work of Chapter 1 to develop and articulate a comprehensive theory of the lawful forest. This next section will introduce the key provocations of the book, provide an opening critique of property's orthodoxy, and contextualise this critique within a broader political economic discourse.

Since the latter nineteenth century, property's orthodoxy has been one that equates property with private commodity and reduces property rights to detached, abstract relations between persons about things. This orthodoxy has been disrupted (but far from dislodged) by the recent emergence of critical property theory and legal geography. This disruption has been achieved partly by highlighting the relevance of context and place, but also by reaching back to a time, not so long ago, when private property was a novel concept. *The Lawful Forest* sees itself as part of this disruptive conversation, drawing together critical property thinking and the contextual lens of legal geography to contest (private) property's dominance.

In so disrupting, this book explores a series of key provocations and asks how the prevailing modern orthodoxy came to be. It seems unthinkable that prior to the eighteenth century, there was 'no clear and unqualified definition of private property in any legal dictionary or the works of any legal writer'.[107] Yet, today, *property* and *private property* appear synonymous. How did this happen? And, why have we forgotten what came before? Why has our 'memory of the commons', our understanding of this earlier, relational understanding of people, property and space, been relegated to the margins, lost to plain sight? Through diverse examples set across history and geography, *The Lawful Forest* identifies the ways in which this residual memory may be unearthed and, in so doing, imagines the radical implications that such an unearthing may invoke.

Property's orthodoxy has been a project centuries in the making, a disembedded edifice built on a skewed – and partisan – historical record, a collusion of capital and power. Moreover, this orthodoxy is domineering, relegating non-conforming spatial relationships, such as the commons, to the periphery. As Blomley explains, private property is so 'universalizing and totalizing' that it renders other forms of ownership 'invisible'.[108] Blomley's 1994 book, *Law, Space, and the Geographies of Power*, is foundational to (what was then) the unnamed field of legal geography. Blomley drew on two

---

[107] R B Manning, cited in Nicholas Blomley, 'Making Private Property: Enclosure, Common Right and the Work of Hedges' (2007) 18 *Rural History* 1, 4.

[108] Nicholas Blomley, *Unsettling the City: Urban Land and the Politics of Property* (Routledge, 2004).

parallel yet converging narratives to explain how law and geography became increasingly homogenised from the seventeenth century onwards. One tale is of *cartography*, how an increasingly sophisticated (and abstract) survey dephysicalised the cadaster from context.[109] The other involves the rise of jurist Edward Coke (1552–1634), who 'constructed a powerful historical geography of the common law'.[110] Coke's mission was to systemise the common law and centralise its judicial administration. The concomitant of this process was the subordination and delegitimisation of place-based custom. Coke wanted to create a unitary legal knowledge throughout England, a universal and 'rational' common law removed from 'the vagaries of social contingency' and 'the multiple legal sites in which the law acquires meaning'.[111]

Blomley argues that Coke crafted '[a] centralized common law [that] was made, *and made quite recently*'[112] – in the process 'effac[ing] an older imagining'[113] and 'shift[ing] the spatiality of legal knowledge' from the 'fragmented and localized' to the 'unitary and centralized'.[114] Of course, Coke's legal geographies were also contested. While his 'appeal to the essential birth-rights and liberties of the free Englishman struck a chord with the propertied', others saw a very different historical geography. As later chapters will canvass, the institutionalisation of a unitary common law defied what Blomley terms the 'folk memories of localized justice', memories held by radical groups such as the Diggers, who disputed the a-contextuality of property and its removal from the commons.[115]

For similar reasons, the enclosure movement is seminal. Enclosure is often seen as a defined historical epoch, which at its height in the eighteenth century extinguished vast swathes of common tenures in rural England. Enclosure was a pivotal moment in property; a schism that alienated millions from their lands, created a class of urban poor available for labour exploitation, and entrenched a private spatial dominance. Enclosure's proponents manipulated property's discourse to advance their self-interest, highlighting the 'improvement' and 'productive use' of 'surplus wastes'. Yet what was 'improved' – and lost to plain sight – was a variegated, diverse and ancient body of localised

---

[109] Blomley attributes this to Christopher Saxton (1542–1610) whose 'politics of mapping' removed the particularities and intricacies of place from the survey, ibid 82–99.

[110] Ibid 71.

[111] Ibid 75.

[112] Blomley, *Law, Space, and Power* (n 67) 4 (emphasis added).

[113] Ibid 77.

[114] Ibid 80.

[115] Nicholas Blomley, 'Law, Property, and the Geography of Violence: The Frontier, the Survey, and the Grid' (2003) 93(1) *Annals of the Association of American Geographers* 121–41.

laws, usages and practices. Today, enclosure has escaped its rural provenance, a logic that in its materiality and in its idea remains as potent as ever. Brett Christophers, for example, identifies a 'new enclosure' in the UK, which he dates from the neoliberal Thatcher era in 1979. This 'new enclosure' continues unabated, enclosing communal resources at rates far in excess of those at the height of parliamentary enclosures. Now, however, it is public lands, not common rights, that are disappearing into the private vortex.[116]

The shift to the city and the capitalist-industrial age likewise changed and shifted conceptions of property. In ways that mirror the historical novelty of private property, Eli Zaretsky argues that 'the rise of industrial capitalism . . . for the first time, [has] created the modern division between the public and the private'.[117] Once removed from a (predominantly) agrarian context, private capital's insatiable demands required an ever more efficient exploitation of land freed from social obligation. As such, land became a commodity, not a locale of relational belonging. By the end of the nineteenth century, property theorists obliged by devising the metaphor of the 'bundle of sticks' – a ruse that reimagined property as a series of detachable, 'go anywhere', discrete rights.[118] By the mid-twentieth century, the right to exclude found itself at the top of this rights woodpile.[119] After all, exclusion best optimised the exploitation of land without undue interference from others. The perfect storm of abstraction, exclusion and the enclosure of diversity has meant that property and its laws now regard 'space as something to be planned over, built on, cultivated, bought, [or] sold . . . a blank canvas . . . to be smoothly acted upon'.[120] Property likewise treats 'its subjects as if they live lives free from context'.[121] Such is this a-contextual, commoditised orthodoxy that has erased an earlier imagining.

And lastly, colonialism was, and is, integral to capitalism. As Rosa Luxemburg emphasised in 1913:

> At the time of primitive accumulation, i.e. at the end of the Middle Ages, when the history of capitalism in Europe began, and right into the nineteenth century, dispossessing the peasants in England and on the Continent

[116] Brett Christophers, *The New Enclosure: The Appropriation of Public Land in Neoliberal Britain* (Verso, 2018).
[117] Eli Zaretsky, *Why America Needs a Left: A Historical Argument* (Polity, 2012) 155.
[118] Stuart Banner, *American Property: A History of How, Why, and What We Own* (Harvard University Press, 2011).
[119] See, e.g., the scholarship of Thomas Merrill or Henry Smith.
[120] Sarah Keenan, *Subversive Property: Law and the Production of Spaces of Belonging* (Routledge, 2015) 21–2.
[121] Ibid.

was the most striking weapon in the large-scale transformation of means of production and labour power into capital. Yet capital in power performs the same task even today, and on an evenmore important scale—by modern colonial policy.[122]

To illustrate her point, Luxemburg highlighted the distinctive difference between the English colonial occupation of India and the many occupations that preceded it:

> The ancient economic organisations of the Indians—the communist village community—had been preserved in their various forms throughout thousands of years, in spite of all the political disturbances during their long history. . . . Then came the British—and the blight of capitalist civilisation succeeded in disrupting the entire social organisation of the people; it achieved in a short time what thousands of years . . . had failed to accomplish. The ultimate purpose of British capital was to possess itself of the very basis of existence of the Indian community: the land. This end was served above all by the fiction, always popular with European colonisers, that all the land of a colony belongs to the political ruler. In retrospect, the British endowed the Moghul and his governors with private ownership of the whole of India, in order to 'legalise' their succession.[123]

Harvey expands on Luxemburg, arguing that colonialism (and other forms of imperialism) is a crucial spatio-temporal fix to capitalism's crises of both underconsumption and overaccumulation.[124] Colonialism helps to stabilise the system, opening up new markets (thus increasing demand for goods) and 'access to cheaper inputs' – 'cheaper labour power, raw materials, low-cost land, and the like'.[125] He highlights Hannah Arendt's argument that capitalism's need for endless accumulation requires the endless accumulation of political power, and elaborates:

> The processes that Marx, following Adam Smith, referred to as 'primitive' or 'original' accumulation constitute, in Arendt's view, an important and continuing force in the historical geography of capital accumulation through imperialism. As in the case of labour supply, capitalism always requires a fund of assets outside of itself if it is to confront and circumvent pressures of overaccumulation. If those assets, such as empty land or new

---

[122] Rosa Luxemburg, *The Accumulation of Capital* (1913) trans. Agnes Schwarzschild (Routledge, 2003) 350.

[123] Ibid 351–2.

[124] Harvey, *New Imperialism* (n 4) 139.

[125] Ibid.

raw material sources, do not lie to hand, then capitalism must somehow produce them.[126]

In Detroit, just as in Sydney and Melbourne, where colonised lands were stolen from their First Nations owners,[127] the implications of enclosure are everywhere on the ground. Owners unable to pay their mortgages, and facing the hopelessness of debt exceeding equity, or crippled by soaring land taxes or water charges, walked away from their homes, leaving a vacuum that private property could not enclose. At the other end of the spectrum, potential owners in Sydney and Melbourne have had to walk away from their dreams as gentrification has priced them out of their communities – or, indeed, out of housing entirely. The inevitable bank foreclosures that followed Detroit's collapse alienated former owners from property's promise: of material security, some measure of wealth, and the proprietarian capacity 'to thrive, to exercise autonomy, to live a human life'. Some abandoned structures, those not demolished, became Detroit's notorious $500 houses; real estate rendered worthless – but for a nominal sum – by private property turning on its own.[128] All, of course, outside the 7.2, or the homogenous gentrified suburbs of Australia's inner-urban belt.

An Outline of *The Lawful Forest*

The Lawful Forest journeys across history and geography, from the forests of pre-Norman England to the urban forests of the modern city, and to the existentialist crises we face on the brink of Gibson's 'Jackpot'. The book stretches from the dawn of common law 'legal memory' to memories fresh in their vivid (and recent) detail. Through its selected, diverse narratives, it engages with praxis, performance and theory.

Chapter 1 lays the foundations for this book's spatio-temporal journey by articulating a theory of the lawful forest. There is an oft-used aphorism that decries our inability to see the forest for its trees, an acknowledgement of the myopic risk of focusing on the minutiae. In this truism, the trope-like use of the 'forest' is no accident. Forests subsist ecologically as interconnected, relational wholes, not atomised collections of their parts. In this way, the chapter seeks to cohere a theory of the forest that speaks to its communal whole – not its individual trees. By building on the Introduction's opening

[126] Ibid 143.
[127] In Australia, the ruse was *terra nullius*. In the United States, lands were taken through treaty stealth, Stuart Banner, *How the Indians Lost Their Land: Law and Power on the Frontier* (Harvard University Press, 2005).
[128] Philp (n 28).

theoretical observations, Chapter 1 suggests alternate ways to better *see* the relational laws and *other* spatial practices that subsist amongst us, glimpsed in activities as diverse as winter solstice festivities or the heat of protest.

Chapter 1 then considers the significance of the space/place dialectic to the lawful forest. Preferring the perspective of critical geography, the chapter argues that the lawful forest is most vivid in moments of disruption, when across the linear grain 'simultaneities' periodically intervene and we are drawn to 'a whole world of comparable instances'.[129] Critical space is never fixed or inert, but rather comprises what Doreen Massey terms a 'simultaneity of stories-so-far'.[130] Meanwhile, critical place is simply a convergence of the momentary 'here and now'. Both accounts – of critical space and place – enhance the rich potential of the lawful forest and its theoretical premises to prefigure a better future.

The chapter ends with the 'phenomenal tree' and its significance for our urbanised, increasingly abstract society. *Things* tend to mediate between the forest and us, distractions that tend to make context disappear. On the one hand, phenomenal trees shatter this artifice by force of their physical mate-riality, the awe-inspiring *thing*-ness of the English oak, Californian coastal redwood, or Australian river red gum. Conversely, the phenomenal tree has become a proxy for the wider forest's loss. This section explores this tension from the perspective of the imperilled Californian redwood and its precarious habitat, threatened by climate change and mega-wildfires.

Chapter 2 then tells a story of the lawful forest that is a millennium or more old, beginning in pre-Norman England, a polyglot patchwork of localised tenures, feudal, allodial and customary. There, property practices reflected the plurality and context of local custom. The Norman Conquest of the eleventh century purported to replace this diversity with uniformity, yet, on the ground, ancient custom proved durable and enduring.

The most egregious example of Norman enclosure was the Conqueror's Forest Law, with the Assize of the Forest of 1189 its apex. Deeply unpopular, this Forest Law privileged royal hunting rights and sought to evict forest residents. Such overreach provoked the Forest Charter of 1217, a statute remarkable for its preference for the forest's 'good and lawful men' over the lay and ecclesiastically powerful. Its 'liberties of the forest' were also remarkable for their longevity, the near-forgotten Charter itself remaining in force for over 750 years. This chapter sets the stage for later chapters, and the dispersal – diaspora-like – of the Charter's 'forest liberties' to the rural commons and beyond.

---

[129] Soja (n 8) 23.
[130] Doreen Massey, *For Space* (SAGE, 2005) 9.

Chapter 3 continues in England, moving into what Marx describes as the 'golden age of labour emancipating itself' – the fourteenth to sixteenth centuries. It provides a historical account of a barely glimpsed utopia, beginning with the shift away from manorialism in the fourteenth century, which culminated in the Peasants' Revolt of 1381. Although this rebellion was short lived, David Rollison argues that a new kind of identity and constitutional norm was forged – a new foundational expectation that those in power would serve the 'commonweal' and be answerable to the people should they fail to do so.[131]

The Revolt was followed by a window of history during which commoners experienced a dramatic increase in freedom, power and shared wealth, before enclosure began to reimpose a reconfigured hierarchical order. In tracing these early enclosures of the commons, this chapter also examines contemporaneous resistance to enclosure through petitions, litigation, direct action and riots. This resistance represented two intersecting narratives, or contests. The first focused on the role of the common people in the government of the day and their broader citizenship rights in the life of the city. The second was specifically around claims to property – and, more significantly, around the meaning and purpose of property.

Finally, Chapter 3 considers another response to this early wave of enclosures in Thomas More's *Utopia* – a seminal work of fiction published in 1516 that both critiqued the times and portrayed a place (or rather 'no place') where enclosure was unthinkable and communal ownership of property the norm.[132] Through the literary arguments of More (in addition to the demands of the Peasants' Revolt and anti-enclosure rioters) this chapter provides a glimpse of an emerging idea of *utopia* – an *other* way of relating to space and community.

Chapter 4 focuses on the more concrete utopias that emerged from the mid-sixteenth century and continue into the present day. This chapter begins with Kett's Rebellion and its camp under the Oak of Reformation at

---

[131] David Rollison, 'The Specter of the Commonalty: Class Struggle and the Commonweal in England before the Atlantic World' (2006) 63(2) *The William and Mary Quarterly – Class and Early America* 221; David Rollison, *A Commonwealth of the People: Popular Politics and England's Long Social Revolution, 1066–1649* (Cambridge University Press, 2010).

[132] Bruce Mazlish, 'A Tale of Two Enclosures: Self and Society as a Setting for Utopias' (2003) 20(1) *Theory, Culture & Society* 43, 45, citing Thomas More, *The Complete Works of St. Thomas More. Volume 4: Utopia*, ed. E Surtz, SJ and J H Hextor; B R Goodey, 'Mapping "Utopia": A Comment on the Geography of Sir Thomas More' (1970) 60(1) *Geographical Review* 15.

Mousehold Heath, outside Norwich in 1549,[133] which has been described as 'the greatest practical utopian project of Tudor England and the greatest anti-capitalist rising in English history'.[134] From the Oak of Reformation, it moves to the Diggers commune, which emerged in 1649 during the upheaval of the English Civil War(s) of 1642–51.[135] The grounded utopians of the Diggers movement established several communities on common land and wastes.[136] In this unsettled period after the execution of Charles I, the Diggers sought to establish a new community – a utopia – in which the commons would be claimed 'for and in behalf of all the poor oppressed people of England and the whole world' and communal farming would create a society free from exploitation.[137] Echoing claims made in the lead-up to the Forest Charter, the intellectual leader of the Diggers, Gerrard Winstanley, argued that this common 'land had been stolen from the people and "hedged into Inclosures" by the rich while the poor lived in "miserable poverty"'.[138] Within eighteen months, all of the Digger communities had been evicted through a combination of legal action and physical intervention.[139] Nonetheless, the Diggers' vision for a new society lives on through Winstanley's writing.[140]

In this chapter, the notion of utopia is explored and defined through historic example and theory. It is used to explain the prefigurative politics of forests, and the experimental communities and protest camps considered in later chapters. Echoing themes raised by MUFI in twenty-first-century Detroit, the discussion of the Diggers also highlights the disputed nature of 'the commons' – due to tensions between bounded and open understandings of community and belonging. It is also in these universalist claims to the commons and the rights associated with them that the inherent links between property and human rights become clear. The language of rights, used by the

---

[133] Ben Maddison, 'Radical Commons Discourse and the Challenges of Colonialism' (2010) (108) *Radical History Review* 29, 33.

[134] Jim Holstun, 'Utopia Pre-empted: Kett's Rebellion, Commoning, and the Hysterical Sublime' (2008) 16 *Historical Materialism* 3–53, 5.

[135] See, e.g., Maddison (n 133); Briony McDonagh and Carl J Griffin, 'Occupy! Historical Geographies of Property, Protest and the Commons, 1500–1850' (2016) 53 *Journal of Historical Geography* 1.

[136] McDonagh and Griffin (n 135) 5; Ariel Hessayon and John Gurney, 'Brave Community: The Digger Movement in the English Revolution' (2008) 59(4) *The Journal of Ecclesiastical History* 778.

[137] Maddison (n 133) 5.

[138] McDonagh and Griffin (n 135) 5, citing 'The True Levellers Standard Advanced', in T Corns, A Hughes and D Loewenstein (eds) *The Complete Works of Gerrard Winstanley* (Oxford University Press, 2009), vol. 11, 5, 13.

[139] Maddison (n 133) 37; McDonagh and Griffin (n 135) 6.

[140] Maddison (n 133) 29–30.

Diggers when they referred to the commons as their 'creation birthright', challenged both the established feudal social order (and its exclusion of the landless) and the capitalist system that was emerging.[141]

From the Diggers, Chapter 4 moves from the rural commons to the city and the 1871 Paris Commune – a 'dramatic seizure of the government by Parisian workers [as] a response to smouldering class antagonisms'.[142] Significantly, this seventy-three-day socialist occupation took place at the beginning of what is known as 'the gilded age', a period that followed a seventy-year increase in income inequality that was about to kick into overdrive. Alexander Vasudevan characterises the Paris Commune as an 'attempt to produce an autonomous social space',[143] while Marx called it 'a resumption by the people for the people, of its own social life'.[144]

According to Henri Lefebvre, the Paris Commune was a 'festive revolution' that created a 'concrete utopia', despite its ultimate demise.[145] In the lead-up to the 1968 uprising in Paris, Lefebvre coined the concept of 'the right to the city'.[146] He described the act of claiming this right as 'autogestion' – a process of seizing control from below in order to collectively manage decisions and common resources.[147] Echoing Winstanley, Mark Purcell argues the right to the city 'is a radical attack on the foundations of capitalist social relations in which the bourgeoisie controls, through private ownership, the means of production'.[148]

Chapter 5 chronicles a growing sense of alienation from the city, and the emergence of environmental protest movements. Its locus is later twentieth-century Australia. At the height of the counter-cultural Age of Aquarius, waves of tertiary-educated youth escaped urban life, seeking radical alternative lifestyles in rural intentional communities. This movement was about a 'return to the land' and its green forests, and the forging of new relationships.[149]

---

[141] Maddison (n 133) 36–8; McDonagh and Griffin (n 135) 6.

[142] Alexander Vasudevan, 'The Autonomous City: Towards a Critical Geography of Occupation' (2015) 39(3) *Progress in Human Geography* 316, 321.

[143] Ibid.

[144] Karl Marx, 'First Draft of *The Civil War in France*', in K Marx and F Engels, *Collected Works. Volume 22: Marx and Engels 1870–1871* (International Publishers, 1986) 486.

[145] Chris Butler, *Henri Lefebvre, Spatial Politics, Everyday Life and the Right to the City* (Routledge, 2012) 35.

[146] Ibid 34–5.

[147] Mark Purcell, 'Possible Worlds: Henri Lefebvre and the Right to the City' (2013) 36(1) *Journal of Urban Affairs* 147.

[148] Ibid.

[149] Benjamin Zablocki, *Alienation and Charisma: A Study of Contemporary American Communes* (Free Press, 1980).

Theirs was an ecological commons, one where 'the city is hardly needed',[150] and where land custodians, not owners, 'repaired the ravages of previous land use battles, and lived in accord with the natural environment'.[151]

The Aquarian communards were at the vanguard of forest protests in north-eastern Australia in the late 1970s. The flashpoint involved plans by the NSW Forestry Commission to log old-growth rainforest in the Nightcap Ranges. Resistance culminated in the four-week Terania Creek blockade beginning in August 1979. Regarded as revolutionising Australian forest protest through principles of non-violent direct action, the blockade set a precedent for practices of lawful dissent across Australia, and has been depicted as the departure point for a new and distinctly Australian protest culture of environmental activism.[152]

Also in the late 1970s, the Hydro-Electric Commission of Tasmania sought to build a dam on the Gordon River, a few kilometres downstream from its junction with the iconic Franklin River. In response, an environmental protection campaign was created by the recently formed Tasmanian Wilderness Society. The ultimately successful campaign was to become 'a seminal political event in late twentieth-century Tasmanian, indeed Australian, history'.[153]

A key player in the Franklin blockade was Wilderness Society founder Bob Brown, who went on to become a long-serving senator and leader of the Australian Greens. In 2016, Brown was arrested under the Workplaces (Protection from Protestors) Act 2014 (Tas.) ('the Protestors Act') for entering 'the Lapoinya Forest for the purpose of raising public and political awareness about the logging of the forest and voicing protest to it'.[154] Brown, and a fellow protester, went on to successfully challenge the validity of the Protesters Act in the High Court of Australia.

While the Court's judgment in *Brown* represents a powerful vindication of the right of peaceful protest, it also appears to validate the ongoing legal significance of the lawful forest. In *Brown*, the Court states that public rights

---

[150] Tim Jones, *A Hard-Won Freedom: Alternative Communities in New Zealand* (Hodder & Stoughton, 1975). Over the next forty years, a similar zeitgeist swept the city, with urban eco-villages proliferating and informing a popular cultural shift.

[151] John Page, 'Common Property and the Age of Aquarius' (2010) 19(2) *Griffith Law Review* 172.

[152] Vanessa Bible, *Terania Creek and the Forging of Modern Environmental Activism* (Palgrave, 2018).

[153] Ian Terry, 'A Matter of Values: Stories from the Franklin River Blockade, 1982–83' (2013) 60(1) *Tasmanian Historical Research Association Papers and Proceedings* 48, 48.

[154] *Brown and another v State of Tasmania* [2017] HCA 43, [14]–[15].

of access to public forests have been long 'recognised' in Tasmanian law. But this 'recognition' indicates that the historic origin of these rights stretches further back – perhaps all the way to the Forest Charter of 1217 and across the long history of the lawful forest.[155]

Chapter 6 concludes by returning to where (and when) our story of the lawful forest began, in our present age of precarity; of climate crisis, global pandemic, structural inequalities and political dysfunction. The core lesson of our critical history of property, protest and spatial justice is to place into proportionate context these past times of societal inflection, periods of rupture when enclosure's linear 'progress' appeared to waver. Taking the longer, critical view reveals that these inflection points were never one-off or aberrational. Rather, these were periodic moments of rich potential, recurring *continuities* that prefigure that the taking of *other* spatial paths could still be within reach. Critically, these continuities reveal that our present (perilous) trajectory is neither preordained nor inevitable.

[155] Cristy Clark and John Page, 'Of Protest, the Commons and Customary Public Rights: An Ancient Tale of the Lawful Forest' (2019) 42 *UNSW Law Journal* 26.

1

A Theory of the Forest

REDWOOD

This book explores an *other* way of relating to land; a relationship to space and place that runs counter to property's (private) orthodoxy and its atomising, exclusionary ways. As set out in the Introduction, we use the metaphor of the *lawful forest* to describe this relationship. Like the forest, this relationship is ancient. Like the forest, this relational understanding of people, place and community seems peripheral to our modern, mostly urban lives, a wooded hideaway beyond our everyday experiences of concrete and steel. While the lawful forest is indeed extant, a literal place of trees, its presence also transcends the physical, evoking a figurative yearning for a spatial life lived better. Simultaneously physical and metaphysical, the forest emplaces us, creating contexts that weave together its remnant presents with a memory of forests past. We also describe this forest as *lawful*, drawing on critical property theorists and scholars of legal geography to derive its ground-up legitimacy from the many, not the few. Aspirations of common-wealth and spatial justice dwell in its shade, a space and place where the 'wide-open synergies'[1] of the commons is a stark foil to the sequestering, controlling inclinations of private capital (and its private property alter ego). This chapter's ambitious task is to situate this vast, diverse (and mostly hidden) lawful forest in theory, and to explain why it remains so peripheral, and so overlooked.

What is remarkable about this lawful forest is its ubiquity. Equally remarkable is its near-invisibility. These observations are, of course, related. Making plain the latter renders obvious the former. It is a forest lost to plain sight, ancient woods that have seemingly disappeared amidst their (individual) trees. This chapter's task (and, indeed, the task of this book) is to reverse this skewed myopia, to identify *why* we see space in reverse, and to advance a theory of the lawful forest that marks its subtle outlines. Or, to imbue this task with a literary flourish, to reveal the forest's concealed glades

---

[1] Carol Rose, 'Romans, Roads and Romantic Creators: Traditions of Public Property in the Information Age' (2003) 66 *Law & Contemporary Problems* 89–110.

and overlooked wood-pastures that subsist amongst us. In so doing, this task requires patience. It takes time to adjust one's vision from the harsh sunlight of the cleared private plains to the forest's dappled shadows. The rewards of patience lie in nuance and connectivity; subtle glimpses of the intricate relational practices and ancient propertied ways that this accustomed sight slowly reveals. Such sightings are the rich bounty of what Vanessa Berry calls the 'radical potential of taking notice'.[2] Such sightings inform and lay the foundations for our theory of the lawful forest.

This chapter's structure, like the forest, tends to a messy interconnectivity, where one theoretical perspective blends porously into the next, a tangled traverse of dense under-stories and overlapping over-stories. The chapter begins by examining the paradox of the forest's near-invisibility, a consequence of design more than accident. It explores how a structural hardening of property has made the *other* disappear, and how language and rhetoric have been employed to entrench this worldview. It next canvasses an eclectic collection of counter-perspectives, which singly and collectively optimise a clearer *seeing* of the lawful forest beyond the single tree.[3] It then turns to a critical (re)interpretation of *space* and *place*, a dialectic replete with implications for our lawful forest. As a *space*, the lawful forest dwells, as noted, in physical and metaphysical form. While long felled to the margins, the tangible forest lingers, not only as enclaves of national parks and federal forests, but also in the edible street verges we plant, the urban wastes we green, and the depleted landscapes we rewild. In its metaphysical iteration, its *idea* reminds us of past struggles for spatial justice and shines a light on the promise of an *other* propertied future. Indeed, its physical/metaphysical duality is unsurprising, since at law, the forest was always bigger than its trees, a legal construct that spoke more to *jurisdiction* than timber, leaf and sylvan meadow.[4]

The lawful forest's spatial turn also disrupts (and at times subverts) what we presume as modern-day legal orthodoxies. Quoting Sarah Keenan (in another context), it shows how property's *other* 'can be productive of spaces . . . shaped differently from networks of hegemonic power relations . . . [and the possibilities of such spaces] creat[ing] new hegemonies'.[5] In taking

---

[2] Vanessa Berry, *Mirror Sydney: An Atlas of Reflection* (Giramondo Books, 2017).

[3] Carol Rose, 'Seeing Property', in C M Rose (ed.) *Property and Persuasion: Essays on the History, Theory, and Rhetoric of Ownership* (Westview Press, 1994) ('Seeing Property').

[4] Forest law has always been concerned with discrete jurisdictional limits, not the acreage of treed forests.

[5] Sarah Keenan, *Subversive Property: Law and the Production of Spaces of Belonging* (Routledge, 2015) 92 (*Subversive Property*).

a spatially disruptive form, the lawful forest lays bare private property's cognitive dissonance, wherein abstraction and exclusion confront lived experiences of context and relationality. The lawful forest similarly challenges ascendant Hohfieldian logic that deems the object of property as a kind of detached 'it'. Instead, it blurs the subject/object divide, such that the propertied object approximates a quasi-subjectified 'you'.[6] No longer a blank, inert canvas, this forested space is co-constituted, performed, dynamic; what critical geographer Doreen Massey describes as 'a simultaneity of stories-so-far'.[7] As our eyes adjust to the forest's dappled light and its dancing, always changing shadows, these overlooked propertied relationships promise an *other* way of performing and (re)imagining the world[8] – by contextualising its subjects/objects, highlighting the vernacular, and exposing private property for its all-too-cosy relationship with 'fiction, fabrication [and] abstraction'.[9]

After space, this chapter then considers the dialectic's other half, the many *place(s)* of the lawful forest. Despite their vast multiplicity and exponential diversity, the many places of the lawful forest share important common ground. These are highly visual places where local trumps universal, context emplaces abstract theory, and heterogeneity is a given. Frequently, lawful forests face existential threat, or endure under an omnipresent sense of siege. As a consequence, the lawful forest's placed-ness is far from neutral. Rather, it is a site of creative tension and political foment, where resistance, radicalism, and even doses of heady utopianism, enjoy a rich provenance. Its political rawness underscores that this is a forest where contest is never far from the surface.

Finally, this chapter closes with the symbolism of the 'phenomenal tree', the proxy for the disappeared physical forest, which, albeit grand in itself, obscures what lies beyond our awed regard for the ancient oak, majestic redwood, or towering river red gum which stands precariously before us.

---

[6] Margaret Davies, 'Material Subjects and Vital Objects – Prefiguring Property and Rights for an Entangled World' (2016) 22 *Australian Journal of Human Rights* 37 ('Prefiguring Property').

[7] Doreen Massey, *For Space* (SAGE, 2005) 9 (*For Space*).

[8] Nicholas Blomley, 'Performing Property, Making the World' (2013) 27 *Canadian Journal of Law and Jurisprudence* 1 ('Performing Property').

[9] Margaret Davies, 'The Consciousness of Trees' (2015) 27 *Law & Literature* 217, 231 ('Consciousness of Trees').

The Pressing Need for Theory

In engaging with a theory of the lawful forest, it is helpful to set the fundaments up front. Ours is a theory crafted in the 'property talk'[10] of the critical socio-legal, of a socially constructed vision of property summative of our diverse relationships with space and place. It is also a theory that seeks to prefigure the future with an eye to the past, or as Margaret Davies describes, one that 'oscillates between imaginings and practice . . . to support alternative narratives'.[11] It is likewise a theory that needs a certain patience to articulate, working backwards from contemporary spatial relationships (and how they have been normalised) – to older, oft-forgotten understandings that both pre-date and coexist with the dominant orthodoxy. In particular, our theory engages with *land* – with all its limitations, physicality and finitudes. In so focusing, we are aware of land's common law reputation for conservatism. As Carol Rose observes, land carries with it a 'veritable flood of doctrine', a torrent of minutiae that smothers profound questions of 'ownership anxiety'[12] in a wave of abstract estates and arcane real property interests. In our theory of the lawful forest, however, the land tells a different tale – not a mind-numbing recitation of hereditaments, profits a prendre and advowsons,[13] but a narrative of material essentialism. Here, finitude by necessity implicates spatial justice. It is land's *scarcity* and the grievance over its uneven distribution that lies deeply unresolved on the forest floor. This is the tension that intermittently flares – a contest inexorably played out between enclosure's 'logic' on the one side and the communitarian forces pitted against it on the other. Periodically, crisis brings this tension to a head, like when a giant emergent tree falls, and dissent's fecund seeds awake from dormancy to flourish in its wake.

The finitude of land and the injustices of space, or rather how space is distributed, underscores the need for coherent theory to inform praxis. Losing sight of the lawful forest implicates our entitlement – and access – to space and place, and concretises inequality. Like the profound injustice of lacking sufficient space to minimise the risks of coronavirus infection canvassed in the Introduction.[14] Or, the structural inequities that condemn poor

---

[10] Carol Rose, 'Canons of Property Talk, or, Blackstone's Anxiety' (1998) 108 *Yale Law Journal* 601 ('Blackstone's Anxiety').

[11] Davies, 'Prefiguring Property' (n 6) 53.

[12] Rose, 'Blackstone's Anxiety' (n 10).

[13] Advowsons are rights to fill an Anglican benefice. They remain good law in England, and 'rights analogous to advowsons have from time to time come before Australian courts', Peter Butt, *Land Law* (Thomson Reuters, 6th ed., 2010) 520.

[14] In Hong Kong, more than 200,000 of the city's poorest residents occupy an average living

and minority citizens to the worst excesses of the climate emergency, where a dearth of public greenery bakes in a structurally skewed 'urban heat island' in blighted suburbs.[15] And so on, and so on, as spatial entitlements oscillate between bounded histories – and presents – and the prefigurative politics of a utopian, boundless *other*. In developing and describing a theory of the lawful forest, this chapter (and subsequently this book) explains why such landed events transpire. It places them in socio-political context. It highlights those forks in the road, of paths taken, and others not, and the alternative 'shadow revolutions' that never materialised. Importantly, it reminds us of a counter-history to enclosure's (mostly unidirectional) flow, of a pushing back against the private sequestration of space and place and its *landed politics of theft*. These are the brief moments that disrupt what passes as an inexorable linearity, the inconvenient interruptions that dominant narratives exhort us to exceptionalise, or forget. These are the under-stories of the ancient lawful forest, and the nuanced theories that underpin them.

Lost to Plain Sight: Failing to See the Lawful Forest for Its Trees

Modern property has a glaring blind spot when it comes to common and public lands. Or more generally, public rights to space.[16] At the same time, property's social and relational incidents have long been trivialised; a reductive process that diminishes and delegitimises ancient law. Stuck in its ideological bandwidth, property normalises a narrow ordering of landed relations that appears rational, neutral and objective. Nicholas Blomley describes this systemic process as 'universalizing and totalizing',[17] a smoothing over of difference that reifies the private estate and its 'land as commodity' ethos, while disappearing the *other*. Unsurprisingly, exclusion becomes property's *raison d'être*,[18] yielding enclaves of 'property' that nestle amidst a non-propertied void. Nicole Graham coins the term 'lawscape' to describe the dephysicalising

---

area of 48 square metres. These crowded tenements are called 'cages' or 'coffin homes', V Wang and T May, 'In Coffin Homes and Cages, Hong Kong Lockdown Exposes Inequality', *The New York Times* (27 January 2021), available at <https://www.nytimes.com/2021/01/26/world/asia/hong-kong-coronavirus-lockdown-inequality.html>.

[15] Brad Plumer and Nadja Popovich, 'How Decades of Racist Housing Policy Left Neighborhoods Sweltering', *New York Times* (24 August 2020), available at <https://www.nytimes.com/interactive/2020/08/24/climate/racism-redlining-cities-global-warming.html>.

[16] John Page, *Public Property, Law and Society: Owning, Belonging, Connecting in the Public Realm* (Routledge, 2021) (*Public Property*).

[17] Nicholas Blomley, 'Enclosure, Common Right and the Property of the Poor' (2008) 17 *Social & Legal Studies* 311 ('Enclosure').

[18] Thomas Merrill, 'Property and the Right to Exclude' (1998) 77 *Nebraska Law Review* 730.

effects of this paradigm, such that law is detached from place and legal landscapes are stripped of particularity.[19] Rebecca Solnit discerns much the same.
In travelling through the American west, Solnit observes the long stretches of
'nothingness', the 'public lands of trees and steepness' or the 'utter-blankness
of desert-dry lake beds' that punctuate private lands 'useful for growing or
grazing'.[20] Translated to the city, Solnit similarly writes of 'streets [that] are
the spaces left over between buildings'.[21] It is this 'nothingness', these 'spaces
left over', which reveals what we have become acculturated not to see as
'things' of property. Nor to value, since as Solnit intuitively surmises, 'prizing
nothingness is harder than prizing something'.[22]

There are many reasons why our heterogeneous understandings of
property – those outside the private hegemony – have been lost to plain
sight. In small part, it is simply how things have turned out, a generational
forgetting overlaid by an accretion of new spatial normals. But in greater part,
it stems from design, not omission, in what Blomley calls 'an organized logic
of purification'.[23] This was, and continues to be, a 'very conscious project',[24]
whereby hegemonic knowledges discredit, conceal or trivialise rival others.
In terms of property, Blomley explains that 'the tendency is to gloss over the
plurality of "legitimate" claims to, and interests in, land',[25] such that spatial
specificity and local difference are erased 'in the name of an ordered and
apparently cohesive unity':

> The Western legal project is underwritten by an organized forgetting of
> such spaces, given that spatial diversity may affect core principles such as
> the rule of law and legal rationality . . . [Sir Edward Coke's] systemization
> of the English common law, for example, entailed the attempt at the crea-
> tion of a unitary legal map in which diverse local knowledges of the law
> were immediately suspect. Increasingly, legal knowledge is imagined as
> disembodied, true to its own internal logic.[26]

In the common law landscape of England, the enclosure movement
was the epochal powerhouse of property diversity loss, a systemic process

---

[19]   Nicole Graham, *Lawscape, Property, Environment* (Routledge, 2011) (*Lawscape*).
[20]   Rebecca Solnit, *A Field Guide to Getting Lost* (Penguin, 2017) 49–50 (*Field Guide*).
[21]   Rebecca Solnit, *Wanderlust: A History of Walking* (Penguin, 2000) 175.
[22]   Solnit, *Field Guide* (n 20) 50.
[23]   Blomley, 'Enclosure' (n 17) 322.
[24]   Ibid.
[25]   Ibid.
[26]   Nicholas Blomley, 'From "What?" to "So What?"', in N Blomley, D Delaney and R Ford
       (eds) *The Legal Geographies Reader* (Blackwell, 2001) 17–34 ('From "What?"').

whereby common lands were privatised, and common rights of use, such as pasturage, the collection of firewood, or the gathering of wild foods, were extinguished. Common rights varied according to local custom and usage. Jeanette Neeson writes of their diversity: 'in reality, on the ground, the range of common produce was magnificently broad, the uses to which it were put was minutely varied, and the defence of local practice was determined and often successful'.[27] Enclosure purportedly began with the Statute of Merton of 1235, when 'wool was power and sheep were the reason' villagers were evicted from their lands. However, as Chapter 2 argues, the enclosing mindset pre-dated its supposed legislative debut. Enclosure is widely regarded as reaching its apogee in the eighteenth century, and 'ended' with Octavia Hill's Commons Preservation Societies of the late nineteenth century, when influential middle-class Londoners (briefly) turned the tide on what they saw as a disturbing loss of open green space. Enclosure changed the diverse propertied landscapes of rural England into a monotone of privatised estates, where 'each section of the grid was dedicated not to the interests of the community, but to the most profitable use its owner could find for it'.[28] It created a landscape of mass exclusion, where fences, hedges and walls divided the many from the few, and inculcated a pervasive mindset that Nick Hayes calls property's 'cult of exclusion'. And in the end, property itself was transformed:

> When the commons were *particularised* (to use a popular euphemism of the time), that is, divided up and sold, these [common] rights were part of the deal and became the sole right of the owner of the land – rights *in* land became rights *to* land. Property had come to be understood less as a network of relations between community and land, and now simply referred to the land itself. The space without the community.[29]

But, of course, enclosure is only one part of what is a relentlessly privatising package. It may be technically accurate to draw the curtain on enclosure by the late nineteenth century, but as critical geographers such as Brett Christophers point out, enclosure has not stopped.[30] It has simply shifted shape, form and nomenclature. Privatisation, encroachment, financialisation, gentrification – the net effect is remarkably similar.

---

[27] Jeanette Neeson, *Commoners: Common Right, Enclosure and Social Change in England, 1700–1820* (Cambridge University Press, 1993) 313.

[28] Ibid.

[29] Nick Hayes, *The Book of Trespass: Crossing the Lines that Divide Us* (Bloomsbury, 2020) 52.

[30] Brett Christophers, *The New Enclosure: The Appropriation of Public Land in Neoliberal Britain* (Verso, 2018).

As flagged, this mindset's 'internal logic' assuages us that space is objective, its 'air of neutrality . . . the epitome of rational abstraction'.[31] Yet the record suggests otherwise, a counter-history that this book chronicles – from the forests of pre-Norman England to the spatial politics of twenty-first-century America. It is a 'logic' that privileges the few, a calculated redrawing of lines that is partisan, not neutral, and rational only to the extent that its rationale is the unidirectional erosion of common-wealth for the benefit of the private. This is a logic that fuels what David Harvey calls in another context 'accumulation by dispossession'.[32] Or, a brutally efficient performance of Pierre Proudhon's '(private) property is theft'. But as Blomley ripostes, what is stolen is not only sheer acreage, but the overlooked possibilities of the *other*; a loss that is both physical and metaphysical in its fact and in its implications:

> The larceny entails that of the diversity (and perhaps the radical potential) of property. Property offers a rich, multivalent and politically charged vocabulary. Cities are sites in which people live inside the ownership model, but also depart from it. Collective claims to land and space are made. And private property turns out to be a good deal more multivalent, both ethically and analytically, than is supposed.[33]

Property Is Persuasion: Weaponising Words

From a closing-off of property's diversity to the ongoing loss of common and public lands, the law and politics of spatial theft is a central plot of this book. And like any narrative, it is enlivened by its words, and how they are weaponised. As others have observed, private property's rhetoric and language are important clues in understanding why the propertied *other* has disappeared from view. If, as Rose argues, 'property is persuasion', then the private estate is adept at 'yelling this is mine' from its bully pulpit, in the process drowning out the quieter other.[34] After all, who could argue with such beneficial private tales? Property is *productive* and *efficient*; it *rewards* the Lockean sweat of the brow; it *improves* the landed polity. And, of course, private property enjoys constitutional privilege, along with other bedrocks such as life, liberty and the pursuit of happiness. Meanwhile, the sunny side

---

[31] Nicholas Blomley, *Unsettling the City: Urban Land and the Politics of Property* (Routledge, 2004) 8 (*Unsettling the City*).

[32] David Harvey, *The New Imperialism* (Oxford University Press, 2003); David Harvey, 'The "New" Imperialism: Accumulation by Dispossession' (2004) *Socialist Register* 63.

[33] Blomley, *Unsettling the City* (n 31).

[34] Carol Rose, 'Possession as the Origin of Property' (1985) 52 *The University of Chicago Law Review* 73 ('Possession').

of property is matched by a darker flipside, where transgressions with private possession are demonised as trespasses. *To wander* and *to roam* connote vagrancy, 'an implicit connection with moral failings', while those who strayed off public rights of way onto private lands were the original 'deviants'.[35] Property's choices of words set and reinforce what is normatively *proper*, and what is not.[36]

The risk of straying from what is proper in our spatial relationships is the risk of re-enlivening legal pluralism. Pluralism challenges the essence of the 'single text' view of the law, one that 'represses the existence and relative autonomy of competing and conflicting socially constituted visions of legal order'.[37] Hence, non-hegemonic rights, like the ancient common right to glean, were degraded into 'mere practice[s]' characterised by their 'vulgarity' and 'promiscuity'.[38] Custom likewise has an ambivalent place in property jurisprudence, since its very nature invokes the spectre of pluralism. Thus, custom is anachronistic, an ancient throwback to localised laws long superseded by a universalising common law. But its peremptory rejection ignores its (once) centrality. As Hendrik Hartog argues, '[f]rom Coke to Blackstone and beyond, custom was a central category in English jurisprudence . . . a protean term, irreducible to clear or coherent categories.'[39] For those who look more closely, custom endures as an apparent outlier, an amorphous source of property law beyond the black letter of case law or statute. Consider in the United States how custom interacts with the laws of beach access, forming the basis of a state-wide customary right to dry sands on Oregon's Pacific coastline.[40] Or how it justifies communitarian access to wet sands versus the tract-by-tract effect of prescriptive public easements.[41] Hartog notes that custom was the law of the poor. It 'formed both the property and the political practice of poor and laboring classes in England and America . . . one of the normative weapons of the weak against the strong, [and] one of the ways in which power was disciplined and concessions enjoyed'.[42] Little wonder it is demonised, and pushed from plain sight. Private property (and private capital) is best served by the bland single story,

---

[35] Hayes (n 29) 18.
[36] Margaret Davies, *Property: Meaning, Theories, Histories* (Routledge, 2007).
[37] Hendrik Hartog, 'Pigs and Positivism' (1985) *Wisconsin Law Review* 899, 934.
[38] *Steel v Houghton* (1788) 126 ER 32.
[39] Hartog (n 37).
[40] *State ex rel. Thornton v Hay 462 P.2d 671* (1969).
[41] David Bederman, 'The Curious Resurrection of Custom: Beach Access and Judicial Takings' (1996) 96(6) *Columbia Law Review* 1375.
[42] Hartog (n 37) 912.

undistracted by 'the plurality and radical decentralization of [other] legal voices'.[43]

The geography of space, like the law of property, also invites a plurality of narratives. Geographer Tim Cresswell says that space can be read as a series of competing normative texts, one dominant and the other subordinate: 'We can . . . talk of a hierarchy of readings, with favored, normal, accepted readings and discouraged, heretical, abnormal readings.'[44] The readings of private space may enjoy a 'favored' status, with their compelling anecdotes of exclusion, dominion and control. Yet those 'discouraged' spatial stories of the 'heretical' and the subordinate continue in their telling, albeit lurking in the shadowed margins. As the next section of this chapter argues, one must simply look (and listen) harder.

Finally, the resort to the language of *tragedy* is equally powerful, as famously employed in Garrett Hardin's 1968 polemic of the tragedy of the commons.[45] There, so-called 'rational herders' fattened their cattle oblivious to the health of communal pastures. Theirs was a metaphor for the inexorable fate of common property, and its tragic over-grazing that individual actors were clueless to stop. Except that the story is disingenuous, a fallacy that depicts a fictional distortion of English history, where it was the enclosers' sheep that overran the commons in contravention of well-established stinting norms, as discussed in Chapters 3 and 4. This one-sided version also ignores counter-examples where communitarian forests and alpine meadows, or group-run riparian rights are successfully managed.[46] Hardin's tale was an ignoring of a fork in the road, a wilful blindness to *other* spatial path(s) taken. Elinor Ostrom later debunked his skewed story,[47] and won a Nobel Prize for her efforts. Yet notwithstanding her scholarly success, it is Hardin's narrative that persists.[48] Communal property is an oxymoron. The

---

[43] Nicholas Blomley, *Law, Space, and the Geographies of Power* (Guilford Press, 1994) (*Law, Space, and Power*).

[44] Tim Cresswell, *Place: A Short Introduction* (Wiley-Blackwell, 2004) 13.

[45] Garrett Hardin, 'The Tragedy of the Commons' (1968) 162 *Science* 1243.

[46] See, e.g., Siddiqur Osmani, 'Participatory Governance: An Overview of Issues and Evidence', in UNDESA (ed.) *Participatory Governance and the Millennium Development Goals (MDGs)* (United Nations, 2008) 1; Ghazala Mansuri and Vijayendra Rao, *Evaluating Community-Based and Community-Driven Development: A Critical Review of the Evidence* (World Bank, 2003).

[47] Elinor Ostrom, *Governing the Commons: The Evolution of of Institutions for Collective Action* (Cambridge University Press, 1990); Elinor Ostrom and Roy Gardner, 'Coping with Asymmetries in the Commons: Self-Governing Irrigation Systems Can Work' (1993) 7(4) *Journal of Economic Perspectives* 93.

[48] One such narrative that has less traction today was Hardin's promotion of eugenics.

commons are tragic. Property is private, individual, atomised and, above all, alienable. Our relationships to space focus on individual 'castles'[49] and we ignore the spaces left over. Once sociable incidents of property are reduced to arcane customs or quaint folktales. The trajectory set, this once contested story is over time concretised, such that plausible others fall from plain view.

Some Implications of Losing Sight of the (Lawful) Forest

In the end, the tragedy of property is not Hardin's spin. Rather, the tragedy is how we have lost sight of property's *other*, a case of 'less [the commons'] supposed internal failures than its external invisibility'.[50] It is a tragedy that confounds the facts on the ground, a kind of 'anti-geography' that takes cartographic form as 'gaps in maps'.[51] Like the vanishing literal forest, we must relearn to see the lawful forest by way of its apparent absence, the holes in the private logic that fail to account for the 'specificities and bonds of place and community'.[52] Or to translate this imperative to cartographic form, we must reimagine (and redeploy) the long-discarded 'sensuous and tactile' contours of pre-modern maps – the artisanal lines of Tolkienesque parish maps that showcased the forests and great trees amidst us and 'drew the world from within, not without'.[53]

Private property thrives on this detached, outside-in perspective, a mind-set that removes the inconvenience of context, carves space into alienable parcels, and delivers up what it portrays as an unpropertied void ripe for the taking. Blomley counsels that if certain spaces 'do not look like property to us, we have tended to ignore them'.[54] Yet they are not overlooked for their gainful potential. In centuries past, these were the pejorative 'wastes' of the English manorial landscape, easy pickings for later enclosure. In the settler colonial state, designated 'empty' hinterlands were preordained for home-steading, white settlement and nation building. Today, they include Solnit's 'spaces left over', *res nullius* emptied of their propertiness – and their propertied alt-potential. Yet they remain spaces and places rich in their monetised opportunity, new frontiers to encroach upon, enclose, appropriate, privatise, gentrify, and so on.

[49] Joseph Singer, 'The Ownership Society and Takings of Property: Castles, Investments, and Just Obligations' (2006) 30 *Harvard Environmental Law Review* 309.
[50] Blomley, 'Enclosure' (n 17) 322.
[51] Blomley, *Law, Space, and Power* (n 43); Blomley, 'From "What?"' (n 26).
[52] Blomley, *Law, Space, and Power* (n 43) 53.
[53] Blomley, *Unsettling the City* (n 31) 91.
[54] Ibid 8.

This *not seeing* but *seeing* is perverse, since despite the disregard, there is always a clear eye to profit. As Blomley highlights, this 'very conscious project' always has and always will turn on monetised gain.[55] Reducing the 'empty wastes' to discrete cadastral parcels is a ceaseless one-way funnel whereby wealth (and power) is transferred from the poor to the rich, the landless to the landed, or the indigene to the coloniser.[56] It is not so much that these 'wastes' are *empty*, but rather they are progressively *emptied* of the rival other. In efficiently shifting resources and power,[57] both the practices and politics of spatial theft rely on two core strengths of private property, its powers to exclude and alienate.

Exclusion is said to be property's gatekeeper right. As Thomas Merrill proclaims, 'give me a right to exclude others from a resource and you give me property'.[58] Exclusion is seen through fences, hedges, high walls and CCTV cameras.[59] Moreover, exclusion possesses a crude physicality, as Jeremy Waldron's plain truth that 'everything that is done has to be done somewhere' makes clear.[60] In England, where 92 per cent of land and 97 per cent of inland waterways are off-limits, doing anything anywhere is vexed.[61] Like finding space to avoid the pandemic. This is yet another plain truth that lies at the heart of private property's 'cult of exclusion' and the spatial injustice that its fences and walls perpetuate. As Nick Hayes opines:

> Walls look like order; but more often than not a wall stands at the fulcrum of an imbalance in society. Most walls are only necessary as a means of defending the resources of those that have them from those that lack them. In this way, though they present themselves as mechanisms of security, they are tools of oppression . . . They underline the strength of what lies within while simultaneously reinforcing the perceived threat of what lies without. They project power as much as protect it. They guard the territory, conceal it, and at the same time announce its presence.[62]

---

[55] Blomley, 'From "What?"' (n 26) 25.

[56] Practised with equal intensity in the metropole as it was (and is) in the colony, Andrew Buck, *The Making of Australian Property Law* (Federation Press, 2006).

[57] Nick Hayes cites the English working-class historian E P Thompson who wrote of the eighteenth century, 'land remained the index of influence, the plinth on which power was erected', Hayes (n 29) 23.

[58] Merrill (n 18) 730.

[59] John Page, 'Property, Values, and the Empirics of Place' (2019) 28 *Griffith Law Review* 1. Its visual antitheses are stiles, kissing gates and 'please close the gate' signs.

[60] Jeremy Waldron, 'Homeless and the Issue of Freedom' (1992) 39 *UCLA Law Review* 295, 296.

[61] Hayes (n 29) 316.

[62] Ibid 95.

Unlike exclusion, alienation is a nuanced concept, a term with multiple meanings, as flagged in the Introduction. In its technical legal sense, alienation refers to the core right to sell, gift, devise and otherwise dispose of one's interest in land. At common law, property is inherently alienable, such that any limitation on that right is jealously guarded, or contained within limits.[63] Yet *alienation* also speaks of property's modern capacity to alienate us from place, the estrangement or sense of detachment that Graham's dephysicalisation thesis portrays.[64] In truth, private property excels at both versions of alienation, a hyper-alienator on steroids.[65]

Alienation's technical meaning drives the feted celebration of land as commodity, where land booms (and busts) ultimately push land prices forever cyclically upwards. This is the process that alienates the *many* from affluent suburbs, or dilutes generations of working-class belonging in gentrified inner cities.[66] In its generic meaning, alienation produces what social commentators variously describe as 'geographies of nowhere',[67] 'sprawlscapes',[68] or 'landscapes of endlessly repeated form'.[69] These wearying, soul-less spaces depict an all-pervasive monotony whose 'motive force' James Howard Kunstler attributes to 'the exaltation of privacy and the elimination of the public realm'.[70]

Thomas Jefferson's National Land Survey[71] is one example where this duality of alienation plays out. In imposing rectangular grids across the continental United States, Jefferson's survey was driven by reputedly altruistic goals, to 'make land available, efficiently and equitably, for purchase and

---

[63] See, e.g., the Modern Rule Against Perpetuities, which renders void any restraint on alienation beyond eighty years, or the common law's lifetime plus twenty-one years.

[64] Graham, *Lawscape* (n 19).

[65] Analogously, Hanoch Dagan writes, '[t]he absolutist conception of property expresses and reinforces a culture of alienation that underplays the significance of belonging to a community, [and] perceives our membership therein in purely instrumental terms', Hanoch Dagan, 'The Social Responsibility of Ownership' (2006) 92 *Cornell Law Review* 1255, 1260.

[66] P E Moskowitz, *How to Kill a City: Gentrification, Inequality, and the Fight for the Neighborhood* (Bold Type Books, 2018).

[67] James Howard Kunstler, *The Geography of Nowhere: The Rise and Decline of America's Man-Made Landscape* (Simon & Schuster, 1993).

[68] Suzannah Lessard, *The Absent Hand: Reimagining Our American Landscape* (Counterpoint, 2019).

[69] Charles Montgomery, *Happy City: Transforming Our Lives through Urban Design* (Farrar, Straus and Giroux, 2013).

[70] Kunstler (n 67) 13.

[71] This vision was later worked into the Land Ordinance Act of 1785 and then revised as the Land Act in 1796.

ownership by individuals' – a republican 'virtue' denied to Native or Black Americans. Yet the orderly division of space into marketable parcels led to what James Corner describes as a 'ubiquitous and standardised built environment, one that looks the same in New York as it does in Anchorage or Albuquerque'.[72] Eventually, Corner concludes, 'every place, regardless of special characteristics, begins to look and feel alike – neutral, flat and bland . . . Differences [are] blended into the lowest common denominator and finally eradicated.'[73] In England, Tim Dee writes of the ongoing threat to local *placed-ness*, where 'specificities have been dulled, local habitations and names globalised, the instress or haecceity of every street or field driven from common memory'.[74] Dee specifically draws the link between ubiquity and private property, and the latter's domineering and unequal effects:

> When I returned to England . . . I looked out at a world that was shaped and built and understood through private ownership, permanent sites of all kinds, and exclusion. Even the open spaces of the north of England were caught in this net of transformation and class. The walls and hedges marked where land had been worked into fields, and where boundaries had been set up. Bare slopes, moorland and bracken were the result of deforestation and grazing of the farmer's sheep. These were landscapes of social and economic exclusions, the basis and expression of great inequity.[75]

Property uniformity divides the world into simple binaries: public or private; nowhere or somewhere; within or without; place or no-place; private wealth versus common; individual trees or (lawful) forests. *No place* is of course the fictional context of Thomas More's *Utopia*, discussed later in Chapter 3, an epic work that invokes yet another binary, the stark polarities of utopia and dystopia. Yet the reflexive division of the world into arbitrary dualities is itself homogenising, rendering nuance and the subtleties of the middle ground less visible and viable. Even the dualities themselves lose their edge. For example, Suzannah Lessard prefers the term 'atopia' to 'dystopia' to describe the proliferation of ex-urban sprawl, the diffuse conurbations that defy traditional understandings of city, country or metropolis. Like the immoral/amoral distinction, atopias are the 'subversive opposite of place as we had known it',[76] sprawlscapes that escape metro boundaries, bleeding

---

[72] James Corner, *Taking Measures across the American Landscape* (Yale University Press, 1996) 32.

[73] Ibid.

[74] Tim Dee (ed.) *Ground Work: Writings on Places and People* (Jonathan Cape, 2018) 3.

[75] Ibid 52–3.

[76] Lessard (n 68) 236.

into former rural localities to disrupt and weirdly disorient any rudimentary sense of place. Dee acknowledges something similar of our material urban geographies, but adds a qualifier. Places do not always have to be 'pretty' to situate meaning, nor to constitute place. Indeed, the modern prevalence of 'less-than-pretty' highlights why spatial curiosity is so critical, a search for the propertied *other* that is all the more critical where the lived landscape is mostly a 'concrete over roar':

> We now understand that the paved world can be as articulate as the vegetated . . . Modernity has shattered our world like never before, we are more deracinated than ever, but because we feel most places to be *nowhere*, we have learnt that *anywhere* can be a *somewhere*. All of our habitat is relevant, not just the pretty bits.[77]

Of course, Dee is not the first person to note the attachment of people to *ordinary* places. For example, Blomley writes of the 'mean and degraded streets' of Vancouver's Downtown Eastside as a proud 'home' to its residents,[78] while geographer Yi-Fu Tuan observes how small-town residents resent an outsider's criticism of its lack of 'architectural distinction' or ' historical glamor'.[79] After all, '[the hometown's] ugliness does not matter', what matters is the residents' 'experience [of] such a small, familiar world, a world inexhaustibly rich in the complication of ordinary life'.[80] If anywhere can be a deeper, granular somewhere, it shows that the hegemony can be over-seen, that it is feasible to look beyond what is in plain view. As noted in this chapter's outset, this is a task of patience, especially so in the most 'paved' of landscapes. In seeing (anew or afresh) the lawful forest, we should take heed of the scattered writings and thinking that help make visible such hopeful glimpses. The next discussion selectively surveys this thinking.

Counter-Ways of Seeing: Glimpsing the Lawful Forest

This section begins with a brief take of critical law and geography. It then finds more fertile soil in their disciplinary intersection, the emerging field of legal geography. It also flirts with the curiosities of psychogeography and the musings of the observant *flâneur*. It ends at a most paradoxical of concepts,

---

[77] Dee (n 74).
[78] Nicholas Blomley, 'Landscapes of Property', in Blomley et al. (eds) *The Legal Geographies Reader* (n 26) 118, 123; Blomley, *Unsettling the City* (n 31).
[79] Yi-Fu Tuan, *Space and Place: The Perspective of Experience* (University of Minnesota Press, 1977) 144–5.
[80] Ibid 144. The concept of 'topophilia' was famously developed by Tuan to refer to the 'affective bond between people and place'.

the confluence of the global city of London and its forests. The idea that 'London is a forest' speaks to how spatial difference may be discerned in the most paved-over of landscapes.[81] Analogously, it exemplifies and enlivens how it is possible to intuit the lawful forest amidst the metropolis, to see what lies amidst and within the 'concrete over-roar'.[82]

As argued, property's detachment from context means that its scholars (for the most part) play invisible games in a curiously abstract sandpit. Private property thrives on its explicit *no-place-ness*, constituted as it is as relations between persons about things.[83] As Thomas Steinberg pithily notes, 'real estate isn't'.[84] Rather, property aids and abets us 'to reimagine and reinvent what we understand to be the real world'.[85] Recent schools of thought, like progressive property,[86] purport to get some sand between their toes, yet it remains wedded to a liberal idea of (private) property. It is in the emerging body of critical and socio-legal property that there is a concerted effort to centre property's materiality. Scholars such as Margaret Davies, Nicole Graham, Sarah Keenan and Antonia Layard, amongst others, recognise property's essentiality and connectivity, and the increasingly urgent imperative to reconcile property with place, community and the environment. Thus, Graham writes of the need to recognise landscapes as 'complex networks of relationships . . . sites of spatial, temporal and legal connectivities . . . rather than as two separate spheres: human and non-human, culture and nature'.[87] Connections and relationships make plain the artifice of property's detachment from place, such that we can better see complex facts on the ground.

A trailblazing property scholar of this genre was Carol Rose. Writing in 1994, Rose talks of 'the enormous importance of visibility in property'.[88] In certain 'forceful and imperious landscapes', Rose says, property patterns 'hit you in the eye', their striking '*imageability*'[89] shaping not only a propertied

---

[81] Paul Wood, *London Is a Forest* (Hardie Grant, 2019).

[82] Dee (n 74).

[83] Wesley Hohfield, 'Some Fundamental Legal Conceptions as Applied in Legal Reasoning' (1913) 23 *Yale Law Journal* 16.

[84] Thomas Steinberg, *Slide Mountain, or the Folly of Owning Nature* (University of California Press, 1995) 17.

[85] Ibid.

[86] Gregory S Alexander, Eduardo M Peñalver, Joseph W Singer and Laura S Underkuffler, 'A Statement of Progressive Property' (2009) 49 *Cornell Law Review* 743–4.

[87] Nicole Graham, 'Sydney's Drinking Water Catchment: A Legal Geographical Analysis of Coal Mining and Water Security', in T O'Donnell, D F Robinson and J Gillespie (eds) *Legal Geography, Perspectives and Methods* (2019) 201–2.

[88] Rose, 'Seeing Property' (n 3) 267.

[89] Ibid 268 (emphasis added).

public discourse, but also the built lie of the land. Rose backed her claim with two examples, the 'imperious' landscapes of Hawai'i and Chicago, where the intersection of 'private and public property law' and their physical geographies lend 'a kind of special visibility' that over time enables people to 'speak to each other . . . about their relations to place'.[90] There, Rose submits, property's imprint can be indelibly seen, from the high-rise shorelines of Waikiki to the 'as far as the eye can see' grids and public lakefronts of Chicago. Rose goes on to argue that 'vision' is 'an essential part of the rhetorical and persuasive equipment of property',[91] citing a capacity to 'see' property through a variety of mediums: as pictures and maps, metaphors, or stories.[92]

Rose says that property as pictures is the 'most obvious and prosaic way to see property'.[93] Hence, visual markers like fences, gates or keep-out signs help us to see private boundaries. Other sightings of property are mediated through photographs and maps, with cartography the subject of especial scrutiny.[94] For example, Rose says that maps 'bring data together in a single perceptible space', in the process 'yield[ing] unexpected new information' and revealing connections that may otherwise lie unobserved.[95] Yet maps are inherently biased, cartographers being free to choose what to include – and omit. Blomley writes of modern maps imposing an 'alien (and legally charged) cartography upon a pre-existent and coherent set of spatial understandings', making the latter invisible.[96] Paul Carter likewise blames Enlightenment cartography for making lines so 'hard and dry', where 'the rectilinear grid imposed on the earth's surface . . . ha[s] no connection to the lie of the land – and in a sense, no interest in it':[97]

> A description of the world is accounted most authoritative when it contains no trace of the knower . . . Maps do this with their alluringly complete

---

[90] Ibid.

[91] Ibid 272.

[92] Ibid 273–4. Rose also suggests a fourth envisioning, of 'illusory property', such as the seeing of pre-emptive 'rights' to a park bench.

[93] Ibid 274.

[94] See, e.g., Blomley, *Law, Space, and Power* (n 43).

[95] Rose, 'Seeing Property' (n 3) 277. Maps also place information in contextual relief, while any distortions of scale or projection are legitimate contrivances since, 'to be practical, a map cannot coincide point by point with reality', Boaventura de Sousa Santos, 'Law: A Map of Misreading. Toward a Postmodern Conception of Law' (1987) 14(3) *Journal of Law and Society* 279, 282.

[96] Blomley, *Law, Space, and Power* (n 43).

[97] Paul Carter, *Dark Writing: Geography, Performance, Design* (University of Hawai'i Press, 2009) 5. This point can be further illustrated by the ongoing social and political impact of the colonial mapping of Africa, pre-partition India, and Korea at the end of the war.

coastlines and calligraphically consistent ranges and rivers. But so do the designed places of urban planning with their suddenly complete patterns of paths, squares, bridges and roads. Nothing moves in these ideal representations. They are theaters from which the possibility of anything happening has been removed . . . How remarkably silent our graphic descriptions of the world are: no breaking surf is heard in them, no animated conversation, no reports of gunfire or anguished whale song.[98]

Notwithstanding the travails and traps of modern maps, Rose's imaginative depiction of property's *visibility* underscores the institution's 'serious state of denial'[99] of context and relationality. In the end, Rose opines, 'there is an old adage, told of plain people and plain things, what you see is what you get'.[100] Here, Rose cuts to the chase. What we see is indeed what we get, an impoverished, decontextualised envisioning of property. What is implicit in this adage, and what 'imperious' landscapes demonstrate in their striking visuality, is that property has not shrunk before our eyes. Rather, it is the other way around. The only narrowing lies in what we have been systemically conditioned to recognise as property. Rose's challenge is to lift our line of sight.

Critical geographers have arguably had greater impact than their legal colleagues in disrupting this a-spatial orthodoxy.[101] Critical geography reimagines space as a complex, dynamic, co-constitutive 'simultaneity of stories-so-far', daily dramas of moving parts more conducive to being 'seen' than the blankness of an inert canvas.[102] Significantly, David Delaney notes critical geography's impact on making plain the visibility of spatial injustice: 'Critical geography is indispensable for revealing the workings of power that conventional spatial imaginaries obscure and for identifying the whys, hows and wheres of injustice that are otherwise *invisibilized* and legitimized.'[103]

More will be said of critical geography later in this chapter. However, for the lawful forest, it is the overlap of law and geography that yields the

---

[98] Ibid.
[99] Rose, 'Seeing Property' (n 3) 297.
[100] Ibid.
[101] Soja argues, for example, that 'Modern Geography has started to come apart at its seams; unraveling internally . . . the grip of older categories, boundaries and separations is weakening. What was central is now being pushed to the margins', Edward Soja, *Postmodern Geographies: The Reassertion of Space in Critical Social Theory* (Verso, 1989) 60. The same could not be said of private property.
[102] See, e.g., Massey, *For Space* (n 7).
[103] David Delaney, 'Legal Geography II: Discerning Injustice' (2016) 40(2) *Progress in Human Geography* 267, 272 (emphasis added).

richest of contextual confluences, bringing as it does a spatial turn to the law. Legal geography is a field that recognises that 'place and law do not live in separate boxes'.[104] Rather, 'context is everything'.[105] As such, legal geography saw its early potential to 'conjure up a powerful challenge to approaches to law which idealise law's separateness, rationality, and reflexivity, and which portray law as deaf [and we suggest *blind*] to material, physical, spatial, and cultural influences'.[106] This 'interdisciplinary intellectual project' recognises the interconnections between law and space, such that the two are 'conjoined and co-constitutive':[107]

> Legal geographers note that nearly every aspect of law is located, takes place, is in motion, or has some spatial frame of reference. In other words, law is always 'worlded' in some way . . . [T]he *where* of law are not simply the inert sites of the law but are inextricably implicated in *how* law happens.[108]

One way to see this intertwining of the *where* with the *how* of property is through Nicholas Blomley's very visible concept of 'property as performance'.[109] In this envisioning, Blomley uses performative theory to suggest that 'our representations do not simply describe a world . . . but they enact a world into being'.[110] In other words, social realities, like the law of property, are 'a relational effect, not a prior ground, brought into being by the very act of performance itself'.[111] Such practices vary from the 'consciously persuasive' to the 'routinized and quotidian'.[112] In this way, private property is enacted by a vast diversity of performances, from 'humble acts of fence building', 'the cutting of hedges', or 'instructions to children not to cross someone else's lawn' to technical performances of mortgage foreclosure or the registration of land titles.[113] Analogously, 'public property is enacted by

---

[104] Jane Holder and Carolyn Harrison, 'Connecting Law and Geography', in J Holder and C Harrison (eds) *Law and Geography* (Oxford University Press, 2002) 19.

[105] Ibid 1.

[106] Ibid.

[107] Ibid.

[108] Irus Braverman, Nicholas Blomley, David Delaney and Alexandre (Sandy) Kedar, 'Introduction: Expanding the Spaces of Law', in I Braverman, N Blomley, D Delaney and A Kedar (eds) *The Expanding Spaces of Law: A Timely Legal Geography* (Stanford University Press, 2014) 1.

[109] Blomley, 'Performing Property' (n 8) 23.

[110] Ibid 25.

[111] Ibid 32. Nor will just any performance do. Rather, they must be sufficiently *reiterative* and *citational* to sustain an enduring precedent value. One-off failures compound a flawed inverse.

[112] Ibid 33–4.

[113] Ibid 33.

kicking footballs around the park, sunbathing on the beach or hiking in national parks, while common property is enacted by community gardening or swimming in your club's member-only pool'.[114] All of these performances are 'worlded' in their legal-geographic doing, but importantly, they take visible form.[115] We can see property in its real-time enactment.[116] Or, as this book details in subsequent chapters, in its performed histories.

Another disciplinary intersection of interest, albeit a quirkier, borderline scholarly one, is that of psychogeography, which, as its name suggests, represents the crossover of geography and psychology. Coined in the 1950s by French *situationist* Guy Debord, psychogeography is premised on a 'search for new ways of apprehending our urban environment', a search that seeks to 'overcome the "banalisation" by which the everyday experience of our surroundings becomes one of drab monotony'.[117] Psychogeography emphasises the experiential, the perceptive and the intuitive, motivated by a radical politics that aimed to disrupt (and transform) the capitalist city through playful stratagems such as urban driftings.[118] Its quest was to 'reveal the true nature that lies beneath the flux of the everyday city',[119] an experiential journey set in the liminal space between objective 'reality' and (at times incommunicable) subjective intuition.[120] In London, the movement persists as a literary school of thought, where amongst other influences, stories of the city's *ley lines* loom large.[121] Ley lines are the indistinct yet deeply intuitive pathways that joined clearings in the ancient forests of the capital, lines that persist *in situ* amidst the modern metropolis.

Psychogeography, for all its quirkiness, speaks to an intuitive instinct about place. A seeing that requires careful, quiet and patient observation. Luke Bennett and Antonia Layard (separately and together) argue that a

---

[114] John Page, *Property Diversity and Its Implications* (Routledge, 2017) 104 (*Property Diversity*).

[115] Cf. Keenan's increasingly abstract and invisible acts of electronic title conveyancing and registration, Sarah Keenan, 'From Historical Chains to Derivative Futures: Title Registries as Time Machines' (2019) 20(3) *Social & Cultural Geography* 283.

[116] Legal geography is also cited for its focus on power: 'The critical mission of the socio-legal scholarship and critical geography from which legal geography has largely emerged seeks to expose power at work, acting quietly but effectively to author the scene in question', Luke Bennett, 'Towards a Legal Psychogeography: Pragmatism, Affective-Materialism, and the Spatio-legal' (2018) 58 *Revue Géographique de l'Est* 1, 3.

[117] Merlin Coverley, *Psychogeography* (Pocket Essentials, 2010) 13.

[118] For example, the use of a map of London to navigate Paris, 'to reveal constraints and [counter] possibilities', Bennett (n 116) 5.

[119] Coverley (n 117).

[120] Ibid 101.

[121] Ibid 119.

pragmatic 'legal psychogeography' could help to shift these amorphous (or what they term) 'affective geographies of matter'[122] into a more rigorous legal-geographic framework. Such a move could help to overcome a gap in law's relationship to space, the capacity to investigate 'an actor's spatio-legal *subjectivities*'[123] without regard to 'spatial/territorial conventions':[124]

> [What is needed is] a new open-mindedness: an actor-centred interpretive approach which [i]s both attentive to, and capable of, portraying how this sense-making necessitates a constant filtering of myriad stimuli and contexts, in which sometimes . . . a legal frame of reference comes to the fore in an actor's understanding of their situation.[125]

An offshoot of psychogeography is the emerging literature and theory of the 'actor-centred' urban pedestrian.[126] In Sydney, Vanessa Berry brings to written life the harbour city's overlooked, enigmatic *other*, revealed from the vantage point of the uber-curious, deeply observant walker:[127]

> The Sydney I know best is one of undercurrents and weird places, sub-urban mythologies and unusual details. These are the city's marginalia, the overlooked and the odd, the hidden and enigmatic places. They form an alternative city, a mirror or shadow Sydney that has the same shape as its better-known counterpart, but emphasizes its ambiguities and anomalies.[128]

Berry employs what she calls 'the radical potential of taking notice'[129] as her tool of choice, a patient strategy that reveals Sydney's 'atmospheres and ambiences' and gives visual form to 'the forces that change the landscape around us'.[130] In particular, Berry intuits the impacts of private property, and how its 'avaricious mood' conforms the city and entrenches spatial inequality: 'The real estate floodlight has come to illuminate every shadow, so no place is spared reduction to economic value. The language of place becomes the

---

[122] Luke Bennett and Antonia Layard, 'Legal Geography: Becoming Spatial Detectives' (2015) 9(7) *Geography Compass* 406.

[123] Bennett (n 116) 4.

[124] Ibid 5.

[125] Ibid.

[126] See, e.g., Frédéric Gros, *A Philosophy of Walking* (Verso, 2017); Olivia Barr, *A Jurisprudence of Movement: Common Law, Walking, Unsettling Place* (Routledge, 2016).

[127] Berry says that psychogeography 'gave a name to the act of examining the urban environment and connecting to the latent forces within it', Berry (n 2) 8.

[128] Ibid 1–2.

[129] Ibid 210.

[130] Ibid 211.

language of ownership, and increasingly these owners are the wealthy and the established.'[131]

From Rose's scholarly 'seeing property' to Berry's personalised 'radical potential of taking notice', a line can be drawn that joins property's overlooked visuality and the power dynamics that such a 'seeing' brings into sharp relief. Like the institution itself, the imageries of property are diverse, ranging from the hegemonic to the subordinate, the patent to the latent, the monetised to the social, the searing bright line to the shadows. Careful seeing, or a 'creative embrace of incongruity',[132] reveals a rich relational bounty beyond property's 'cult of exclusion', such that we see a fuller picture of property, a glimpse of the lawful forest as well as its single trees.

This section ends with a vignette from the global city of London. It is a reminder of the literal forest amongst us, and the type of careful discernment that may equally reveal its alter-lawful forest. Tim Dee observes of his English village that 'it is a *depthy* place engrained with stories'.[133] Dig beneath the 'concrete over roar' of metropolitan London, and no doubt the depthy stories of its many 'villages' likewise rise to the surface. Indeed, the act of digging is probably superfluous, given this chapter's oft-repeated refrain of things being lost to plain (surface) sight.

Paul Wood dedicates his 2019 book *London Is a Forest*, a light-hearted guide to the metropolis's forest, to 'The Woods'. However, Wood does not prosecute a metaphor. His brief is straightforward; London *is* a literal forest. Using sensuous maps and guided walking trails, and tracing the faint imprints of forest highways, Wood discerns the shape and multiple forms of London the forest. He locates 'boundary trees' still standing amidst the asphalt and traffic, reminding Londoners 'of the time when natural landmarks were important demarcations of borders between landowners, parishes, rights of way, and other jurisdictions'.[134] In the crowded inner city, the forest takes on a 'decidedly urban feel', restricted to 'gardens, parks and the edgelines of railway lines'. Yet despite their sparse cover, these ancient woodland remnants 'provide a glimpse into the past, and maybe, a model for the future'.[135] In high-rise estates, allotments and forest gardens 'foster communities and connect people with the food they eat',[136] while a careful eye spots misshapen tree trunks that bear witness to ancient practices (such as

---

[131] Ibid 12.
[132] Bennett (n 116) 4.
[133] Dee (n 74) 17.
[134] Wood (n 81) 135.
[135] Ibid 35.
[136] Ibid 54.

pollarding and coppicing) once enjoyed as common rights. In the city's deer parks and remnant commons, 'old and sustainable practices continue', or are remembered as ancient fire pits or other telltale physical signs. Elsewhere, the names of streets, suburbs, or stops on the Piccadilly Line, speak to London's forested past.[137] Even when submerged, its ancient forests are never far from sight. At low tide, 'the primeval past can be glimpsed in the preserved tree stumps exposed . . . in the Thames mud at Erith'.[138]

Wood's *London Is a Forest* is not a scholarly work. But its frequent invocations of ancient lawful relations to space, of public rights of way, surviving common rights, the forgotten statutes of nineteenth-century Commons Preservation Societies,[139] even the lingering effects of the Forest Charter of 1217, present as a subtext to this book's lawful forest. Wood's argument is simple, whether in terms of the percentage of tree canopy (20 per cent), or the number of trees per resident, London meets United Nations definitions of an urban 'forest'.[140] Yet, technicalities aside, London's forest demands that it is seen beyond its many trees. His scattered yet holistic sightings of this forest are intuitive performances of a radical taking of notice. To Wood, London as a forest is not lost to plain sight, in its presents, pasts, or imagined, prefigurative futures. Such is the clear-sighted template we all need.

The Spaces of the Lawful Forest

In discarding private property's myopic blinkers, it next helps to know *where* the lawful forest grows and flourishes. After all, knowing where to look makes it easier to see. We argue that the lawful forest subsists in the critical geographies of space and place, sites characterised by openness and flux, materiality and diversity. Here, *space* is dynamic, a 'reference system'[141] constantly made and remade. *Place* is found in its passing spatio-temporal convergences, an immediacy that Massey terms a 'here-and-now throwntogetherness'.[142] And, of course, context is everything. This section explores the implications of this radical space/place duality; highlighting its theoretical richness, while

---

[137] Ibid 133.

[138] Ibid 12.

[139] Wood writes of the lasting reforms of Bermondsey Mayor, Ada Salter, who in 1919 planted an urban forest in her Borough, ibid. Or, the Epping Forest Act of 1878, which along with other like statutes marked the formal legal end of enclosure, Hayes (n 29) 276–7.

[140] 'The city is home to over eight million trees, roughly one for every person, and enough for 20% of the capital to be covered by tree canopy', Wood (n 81) 11.

[141] David Harvey, *Justice, Nature and the Geography of Difference* (Blackwell, 1996).

[142] Massey, *For Space* (n 7) 140.

drawing on the insights of critical and socio-legal scholars whose complementary ideas likewise flourish under the lawful forest's canopy.

The orthodox conception of *space* renders it neutral and inert, a *tabula rasa*. As Keenan explains of the law's relationship with space, 'legal judgments, executive powers, legislation and legal commentaries tend to treat space as something to be planned over, built on, cultivated, bought, sold and/or protected; a blank canvas . . . to be smoothly acted upon'.[143] Blomley writes similarly of the common law as a 'disembedded superstructure' where space 'increasingly appears as an objective and pregiven surface' and where 'the law can only be interpreted at the highest level of *spatial* abstraction'.[144] Indeed, in this worldview, legal abstraction renders space almost 'irrelevant', supervened by a single narrative 'a-spatial language of order, equality, and the homogenous rule of law'.[145] Space approaches a *no-space*, a context so bland that the lawful forest disappears into the all-consuming foreground.

In this iteration, space also presents as linear, a dimension of orderly progress, or at least, one of ineluctable trajectory. Colonialism, for instance, treats certain spaces as 'developed' or 'advanced', while others are further back in the spatio/historical queue.[146] Proponents of enclosure or gentrification argue analogously, citing hierarchies of 'progress' and 'improvement'. Massey's critique of this linear understanding of space lies in its perversity, it is a 'sleight of hand', a 'proposition [that] turns geography into history, *space into time*'.[147] Edward Soja's seminal work, *Postmodern Geographies*, likewise critiques the 'hegemonic historicism' of space that renders what he calls Modern Geography 'theoretically inert'.[148] Soja says that this mindset leads to a 'critical silence' in contexts where 'an overdeveloped historical contextualization of social life and social theory . . . actively submerges and peripheralizes the geographical or spatial imagination'.[149]

By contrast, the lawful forest dwells in this imagining of the spatial *other*. In the lawful forest, space is the antithesis of the conventional 'blank canvas waiting to be smoothly acted upon'. Instead, it is dynamic, performative, relational, co-constituted and inherently social. Critical space also conforms to different 'reference systems', whether physical, conceptual, or social, akin to the lawful forest's physical/metaphysical duality. Keenan says these three

---

[143] Keenan, *Subversive Property* (n 5) 21.
[144] Blomley, *Law, Space, and Power* (n 43) 76, 91 (emphasis added).
[145] Ibid 107.
[146] Massey, *For Space* (n 7).
[147] Ibid 5 (emphasis added).
[148] Soja (n 101) 16.
[149] Ibid 15.

spatial conceptions are interlinked and frequently coalesce; however, she argues there is merit in clarifying each. Physical space refers to our 'clearly tangible environment, whether "natural" or "built"'. Social space is 'the socially and culturally created yet also tangible surrounds in and through which we live', while conceptual space is 'the realm of abstract ideas and designs about how the physical and social world does or should operate'.[150]

Importantly, this critical understanding of space is non-linear, neither preordained in its outcomes, nor pregiven. Soja cites art critic John Berger to illustrate this understanding, who in 1972 argued for 'new ways of seeing' space and time:

> We can no longer depend on a story-line unfolding sequentially, an ever-accumulating history marching straight forward in plot and denouement, for too much is happening against the grain of time, too much is continually traversing the story-line laterally . . . Simultaneities intervene, extending our point of view outward in an infinite number of lines connecting the subject to a whole world of comparable instances, complicating the temporal flow of meaning . . .[151]

Indigenous spatial worldviews are analogously non-linear, collapsing into a merged one-ness. As Mary Graham states, '[t]here is no division between the observing mind and anything else: there is no "external world" to inhabit', such that 'all aspects of existence continually interpenetrate each other'.[152]

In this book, like Berger's (re-)altered perceptions, our storylines 'continuously traverse' the linear, running against the grain. The spaces and places of the lawful forest dwell on these fault lines, where 'simultaneities intervene' and we are drawn to 'a whole world of comparable instances'.[153] These are our tales of the lawful forest, where succeeding chapters interrogate a 'whole world of comparable instances', from Kett's Rebellion to the Paris Commune and to modern-day urban protest camps, and more. These are events particular to their context, yet they are not random occurrences. They recur when particular social, economic, legal and political factors align, 'comparable instances' when the distribution of land proves so egregious that 'people are liable to kick back'.[154] In such heady times, the 'simultaneities of stories so far' share a common plot of a running against enclosure's grain and

---

[150] Keenan, *Subversive Property* (n 5) 11.
[151] Soja (n 101) 23.
[152] Mary Graham, 'Some Thoughts about the Philosophical Underpinnings of Aboriginal Worldviews' (1999) 3 *Worldviews: Environment, Culture, Religion* 105.
[153] Soja (n 101) 23.
[154] Christophers (n 30) 324.

a societal (re)imagining of a spatial *other*. Karl Polanyi called such moments the 'counter-movement'. These were times when 'human societies tend to react forcefully to the privatization of land, to its treatment as a commodity to be bought and sold, and to the negative outcomes that . . . often flow from such treatment'.[155] These are times when the overweening myopia briefly lifts, and the lawful forest is clearer to plain view.[156]

So, how does the space of the lawful forest appear? Hayes provides an unintended example in his description of a contested footpath near his childhood home in rural England. Hayes calls this humble path 'democracy manifested in mud',[157] a depiction that in its simplicity captures the critical spatial characteristics of the lawful forest and highlights the stark contrasts of private space:

> [The path's] direction is determined by the efficiency for the people, its legitimacy by use and engagement. . . . The path that runs from my woods in West Berkshire, up past the dell, through the empty woods to join the Right of Way is one such democratic path. That property law pretends it is not a public path is another brash denial of the 'pure bleeding obvious' the hundreds of thousands of feet that pressed it into being. The properties of the path are entirely contingent on its being the property of the public, in the Lockean sense, by virtue of the work they put into it . . . [The wall around Basildon Park] severed the path, blocked the old route to the river and transformed the function of the land in between. On the will of one man alone, it corralled the commoners of my village like livestock around its newly sanctioned space.[158]

First, Hayes's footpath is *relational*. It enlivens a 'specific conception of the relation between individuals [the local villagers], and the object of property'[159] (their 'rights' to the track site). Its meaning is therefore greater than its efficient transport of villagers from A to B. Rather, its meaning lies in the affinity it engenders, the relationships forged *in situ* by literally centuries of foot passage. Relationality is 'a productive line of thinking' that does not

[155] Ibid 324.
[156] Analogously, protest literature observes how 'protest movements tend to cluster in what are sometimes called "movement eras" [when] disturbances, especially economic disturbances that provoke one group to rise in anger and defiance affect others whose circumstances may be similar', Frances Fox Piven, 'After the Crash: Searching for Alternatives', in J Gantz (ed.) *The Age of Inequality: Corporate America's War on Working People* (Verso, 2017) 210.
[157] Hayes (n 29) 143.
[158] Ibid.
[159] Blomley, 'From "What?"' (n 26).

begin its enquiry 'with the *thing* first and subsequent relations second', but rather recognises the primacy of the 'meaning, significance and identity' that people derive from their relationship to particular space.[160] The concept of villagers *engaging* with the path reflects this relationship, felt acutely in the generational loss of the 'old route to the river',[161] yet still enjoyed (in part) as it winds 'past the dell and through the empty woods'. Relationality depends on intimacy with context (here a specific pathway in West Berkshire), and how that fosters a sense of belonging. *Property as belonging* is a sympathetic school of thought that emphasises a propertied subject's affinity to context. Keenan says that belonging is 'a necessarily *relational* term', one that describes 'the state of fitting smoothly, or without trouble, into a conceptual category or a material position'.[162] Belonging and property are logically related, Keenan argues, because 'spaces where subjects belong are spaces of property'.[163] Davina Cooper reinforces belonging's centrality to property, arguing it is 'the first, and most important, aspect of property practice', one which lies at the 'definitional and normative level' of both 'subject/object and part/whole orientations of property'.[164]

Second, this pathway is enacted and reiterated by its *performance*, by the Lockean work put into it by 'the hundreds of thousands of feet that [literally] pressed it into being'. Performativity has been canvassed earlier in this chapter, a theory that recognises property relations arising by their (qualitative, ever dynamic) acts of doing:

> Performances of property . . . are both citational, referencing numerous other performances, and reiterative, entailing sustained forms of re-performance. [Moreover] property claims are continuously remade and re-enacted, and, as such, open to surprise and complexity, yet also capable of fixity and sedimentation.[165]

Performativity also rebuts the blank spatial view that property is vested in only one or two critical points in time (for example, the creation of the original title, or its subsequent transfer). Rather, time is always relevant,

---

[160] Marie-Eve Sylvestre, Nicholas Blomley and Céline Bellot, *Red Zones: Criminal Law and the Territorial Governance of Marginalized People* (Cambridge University Press, 2020) 27. Sylvestre et al. also observe how a 'territoriality's relationality . . . can easily disappear from view', ibid.

[161] Page, *Public Property* (n 16) and the concept of 'public connection', 103–35.

[162] Keenan, *Subversive Property* (n 5) 12 (emphasis added).

[163] Ibid 14.

[164] Davina Cooper, 'Opening Up Ownership: Community Belonging, Belongings, and the Productive Life of Property' (2007) 32 *Law & Social Inquiry* 625, 629.

[165] Blomley, 'Performing Property' (n 8) 25.

a continuum that in the case of Hayes's pathway runs across centuries of footfalls. Performance theory resonates with the dynamism and ground-up legitimacy of the lawful forest, an ongoing spatial work-in-progress marked by constancy and the citational power of particularly resonant acts of doing.

Analogous to performance theory, the lawful forest is also *co-constituted*; a material space constructed by both the physical/metaphysical forest itself and those who pass through it. Keenan's theory of belonging, for example, is premised on the emplaced subject 'taking space with her as she moves',[166] while, at the same time, (re)shaping the spaces in which she is physically embedded'.[167] In a similar vein, Olivia Barr sees the common law as co-constituted by the juridical movements of those who pass through its jurisdiction:[168] 'To move and walk, for instance, is to materially practise common law and participate in the creation and conduct of lawful relations.'[169] Barr's jurisprudence of movement also challenges the common law's erstwhile linearity:

> We live in a world of excessive linearity, and our understanding of the dynamic place of law suffers as a result. Lines however are not so straightforward. Lines have linings; they have outlines and inlines, texture and detail, they break and cross paths, moving and wandering as they fail to meet.[170]

Barr's image of lines 'breaking', 'crossing' and 'failing to meet' is similar to Berger's, where the vicissitudes of context and the dynamism of human actors passing through conspire to shape the lawful forest's disruptive contours. Hence, in Chapter 4, the Diggers' colony atop St George's Hill in Surrey was as much informed by the Crown wastelands on which it was sited as it was by the philosophies of its radical croppers. Or, in Chapter 5, where countercultural communards established their transitory settlements in the green subtropical forests of 1970s Australia, such that space and subject forged an antipodean dawning of the Age of Aquarius. And of course, in Hayes's West Berkshire village, the pathway is more than a convenient thoroughfare; it is a (broken) conduit of communitarian significance formed over long use.

---

[166] Keenan, *Subversive Property* (n 5) 16.

[167] Ibid 87. Delaney's 'nomosphere' likewise 'provides a way of thinking about the complex, shifting, and always interpretable blendings of words, worlds and happenings in which our lives are always embedded and through which our lives are always unfolding', David Delaney, *The Spatial, the Legal and the Pragmatics of World-Making: Nomospheric Investigations* (Routledge, 2010) 26.

[168] Barr cites the acts of walking and burial in the context of the settler colonial state, Barr (n 126).

[169] For example, prescription is a property law doctrine where rights are acquired through open, not secret, peaceful, non-consensual use over twenty or more years.

[170] Barr (n 126) 13.

Space, context and its emplaced subjects conjointly create the materiality of the lawful forest and its metaphysical, defining *zeitgeist*.

Fourth, the space of the lawful forest is inherently social. For example, the glue of Rose's 'comedic commons'[171] is its underrated sociability. Rose likens public sociability to a scaled return on investment. The greater the public engagement in land, the wider buy-in and the higher its value to individuals participating in the sociable activity.[172] This is the felicitous merry-go-round where 'the more is the merrier' and the 'better for all',[173] a 'buy-in' that works to protect the space from enclosure. And not only is the lawful forest social, its social relations construct its space. Amelia Thorpe observes that successful public spaces 'are grounded in relationships of social and material connection to the site in question'.[174] Such relationships arise through histories of connection to the land, such as residence, employment, social networks or community ties.[175] More widely, in a critical theory sense, they co-constitute a wider public 'ownership':

> Ownership is a type of social practice, one that both reflects and forms social structures. People feel ownership over places where they have established both social and material connections: places they have spent time being physically present and, importantly, engaging with others.[176]

Lawful forests are famously sites of rich sociality. In the rural commons of England, Jeanette Neeson observed the 'intimate spatiality of the classic commons',[177] where traditional uses such as pasturage were complemented by non-instrumentalist rights replete with social meaning and relational implication. Neeson records how 'access to the openness of the commons brought solitude . . . each usage of common waste created a sense of self; it told commoners who they were, a part of a tribe'.[178] Fast-forward to the early twenty-first century, and the camps of Occupy were not only places of anti-capitalist protest or struggles for public space,[179] but also places of social fabric and festive celebration. Bonnie Honig says that the first thing built at

---

[171] Carol Rose, 'The Comedy of the Commons: Custom, Commerce, and Inherently Public Property' (1986) 53(3) *The University of Chicago Law Review* 53, 711.

[172] Page, *Public Property* (n 16) 63–7.

[173] Rose, 'Possession' (n 34) 767–8.

[174] Amelia Thorpe, 'Pop-Up Property: Enacting Ownership from San Francisco to Sydney' (2018) 52 *Law & Society Review* 740, 742.

[175] Page, *Public Property* (n 16).

[176] Thorpe (n 174).

[177] Cited in Blomley, 'Enclosure' (n 17) 320.

[178] Neeson (n 27) 180.

[179] Gantz (n 156).

Occupy Wall Street was a public library.[180] In Occupy Sydney, there were 'musical performances, film nights, and free yoga and meditation classes',[181] while in Gezi Park, Istanbul, occupiers founded a makeshift 'clinic, communal kitchen, nursery, library, communications office and market garden'.[182] These twenty-first-century occupations unwittingly echoed earlier instances, such as the sixteenth-century rebel camp at Mousehold Heath, where Robert Kett and fellow rebel campers ran a thriving camp kitchen and held regular church services.[183]

Last, the lawful forest tends to erode the arbitrary distinction that orthodox theory draws between the subject(s) and the object of property. Wesley Hohfield famously describes property as a series of relations between persons about things. The thing itself is no longer 'property', but simply the object of the juridical relationship.[184] This contested dichotomy lies at the heart of critical property theory. For example, Margaret Davies identifies that forests and (especially) 'phenomenal trees' tend to mediate and disrupt cultural divides that humans have constructed to explain 'their' world; divides of culture and nature, human and non-human, person and property, subject and object. Davies tells the story of an ancient river red gum, enclosed in a massive glass dome in a shopping mall in suburban Adelaide, Australia. Once flourishing in open forested space, the legally protected tree becomes a centrepiece of the mall's redevelopment. However, the radical change of its environment – the tree loses access to natural rainfall, humidity and ultraviolet light – eventually causes its terminal decline and death. The tree refuses to bend to human conceit, such that its resistance and demise becomes symbolic of 'a more dynamic understanding of the subject–object relationship in property, a better understanding, in fact, of the role that "objects" play in the construction of proprietary relationships'.[185] Davies argues that forests (and their trees) are objects with an independent social and propertied identity, 'interpreted and juridified by human intervention'.[186] This perspective enables forests, as 'integral parts of both our survivability and our sociality', to

---

[180] Bonnie Honig, *Public Things: Democracy in Disrepair* (Fordham University Press, 2017) 20.

[181] Cristy Clark and John Page, 'Of Protest, the Commons and Customary Public Rights: An Ancient Tale of the Lawful Forest' (2019) 42 *UNSW Law Journal* 26, 32; *O'Flaherty v City of Sydney Council* (2013) 210 FCR 484, 488 [15] (Katzmann J).

[182] S Erensu and O Karaman, 'The Work of a Few Trees: Gezi, Politics and Space' (2016) *International Journal of Urban and Regional Research* 1, 12.

[183] See Chapter 4.

[184] Hohfield (n 83).

[185] Davies, 'Consciousness of Trees' (n 9) 222.

[186] Ibid.

'achieve a kind of subjectivity' as things 'entangled' in human communities, a complex entanglement that shifts our understanding of the forest and its trees from a detached 'it' to an increasingly subjectified 'you'.[187] As Davies adds, 'it becomes a "you" basically by a dialogical process, it draws a person in, who then responds to its interpellations',[188] much like Christopher Stone's parched, yellowing lawn that spoke of its need for water.[189] In the lawful forest, property's objects tend to become quasi-independent actors in the real-time dramas that unfold in context.

As flagged in this chapter's introduction, spatial theorisations of the lawful forest tend to a messy interconnectivity, where one perspective porously bleeds into another. There are hard to distinguish under-stories and overlapping over-stories. We have seen how closely related sociality and relationality are, and how performativity and co-construction are in substance species of the same genus. The space of the lawful forest possesses all of these attributes, and analogously more. What is next explored is that momentary collision of space and time, the convergences and (occasional) conflagrations that constitute the diverse and many places of the lawful forest.

The Places of the Lawful Forest

If critical space is 'open, multiple and relational, unfinished, and always becoming',[190] then critical *place* is a snapshot of that becoming. While both space and place 'emerge through active material practices',[191] space provides the contextual 'reference system', while place is fleeting. Massey describes a train journey from London to Milton Keynes, in which the *place* of London that the traveller has left is no longer the London of now: 'Lives have pushed ahead, investments and disinvestments have been made in the City, it has begun to rain quite heavily, a crucial meeting has broken up acrimoniously',[192] and so on. And arriving at the new place is to pick up its threads and to 'weave them into a more or less coherent feeling of being "here" [and] "now"'.[193] This all contributes to the specificity of place, in what Massey colloquially terms the 'throwntogetherness' of the 'here-and-now'.[194]

---

[187] Ibid 229.
[188] Ibid 225.
[189] Christopher D Stone, 'Should Trees Have Legal Standing?' (1972) *Southern California Law Review* 450.
[190] Massey, *For Space* (n 7) 59.
[191] Ibid 118.
[192] Ibid.
[193] Ibid 119.
[194] Ibid 140.

Yet place's narratives are not standalone: 'the successions of meetings, the accumulations of weavings and encounters build up a history. It's . . . the very differentiation of temporalities that lend[s] *continuity*.'[195]

The many places of the lawful forest represent a seemingly infinite convergence of countless here and nows. Yet again like the common threads of space, they do reveal *continuities*, an accumulation of pasts and presents in which there are distinctive patterns of place. This section explores some of these recurring patterns.

First, the places of the lawful forest are inherently visual, the lie of their land easier to *see* (in Rose's sense) in their brief moments of incandescence. Stratagems such as the radical potential of notice, affective legal psychogeography, or performance theory need less of a workout in these clear-eyed, bright-line moments. One can easily picture how vivid the Diggers' camps must have been, given their ongoing scholarly depiction centuries after their brief moment in the English sun. There is something deeply visceral in the visuality of the lawful forest's place, a seeing that is faithful to its many recordings.

Next, the lawful forest's place is local, an expression of the vernacular. As canvassed, this conception is heresy to orthodox property and its demand that all propertied relations are reduced (and reducible) to a *universal sameness* in the interests of information cost.[196] However, context is a nagging, inconvenient counter-narrative. As John Orth pithily observes, 'all property law is local . . . the place where the land lies'.[197] Thomas Steinberg says much the same, revealing the absurdist fallacy that 'real estate isn't'. Rather, Steinberg uses local case studies to assert that property 'penetrates everywhere in the realm of daily life'.[198] There is a need to recover the importance of

---

[195] Ibid 139 (emphasis added). Other geographers distinguish space and place using different metaphors. Tim Cresswell says that space is a realm without meaning, but when 'humans invest meaning in a portion of space, and then become attached to it in some way . . . it becomes a place', Cresswell (n 44) 10. Tuan differentiates space and place, with the former being 'open' and the latter 'resting', Tuan (n 79). Blomley says that place is a 'slippery concept', Blomley, *Law, Space, and Power* (n 43) 112.

[196] Thomas Merrill and Henry Smith are seminal proponents of the *numerus clausus* view of property, the closed list of (private) property rights. Unlike contract law, and its capacity to create new and different rights *inter partes*, property law relies on the informational cost imperative of having a closed list of universalised rights. There can be no 'Monday watch' in this worldview, Thomas W Merrill and Henry E Smith, 'Optimal Standardization in the Law of Property: The Numerus Clausus Principle', (2000) 110 *Yale Law Journal*.

[197] John Orth, *Reappraisals in the Law of Property* (Ashgate, 2010) vii.

[198] Steinberg (n 84) 9.

place in property law,[199] and the local-ness of the lawful forest is a reminder of that. If place is space suffused with meaning, as Cresswell suggests, then 'nowhere are the rhetorical foundations of *places* more clearly revealed than in their *local* histories'.[200] Local also inheres diversity,[201] the idiosyncratic and always-heterogeneous patterns of place that vary from context to context. As Joseph Sax posits, 'diversity is a good thing, in human settlements as well as nature'.[202]

The places of the lawful forest are also rich in public wealth, or at least in their aspirations for it. Frequently, such places come out of spaces that are commensurately pauperised by an often-crushing public poverty. In these instances, we see place-based responses as a vigorous pushing back against spatial inequality. Public wealth is an under-theorised concept. It is certainly *material*, in the sense that tangible public places are sites of public wealth. Expansive parks, urban squares, cycle paths, public swimming pools and libraries, these constitute a public infrastructure of 'concrete and steel, grass and greenery'[203] that enhances sociability and promotes happiness. Charles Montgomery makes the link between public resources and happiness, using the city of Bogotá, Colombia, during the mayoralty of Enrique Peñalosa as his case in point.[204] Peñalosa addressed his city's *unhappiness* by an expansion of public wealth, with a rollout of parks, improved public transport and public libraries in poorer suburbs, amongst multiple initiatives. While all this cost money, Peñalosa was driven by a simpler ethos:

> We need to walk, just as birds need to fly. We need to be around other people. We need beauty. We need contact with nature. And most of all, we need not to be excluded. We need to feel some sort of equality.[205]

Public wealth may also be intangible. Many public rights, such as the right of recreational access in Scotland, are incorporeal.[206] Or Monbiot's proposal for an urban right to roam, canvassed in the Introduction, is another incorporeal

---

[199] Analogously, in a case-based study of environmental law, John Nagle observes 'there is a special need . . . to recover the importance of place in environmental law', John Nagle, *Law's Environment: How the Law Shapes the Places We Live* (Yale University Press, 2010) 252.

[200] Carter (n 97) 122.

[201] Page, *Property Diversity* (n 114).

[202] Joseph Sax, 'Do Communities Have Rights? The National Parks as a Laboratory of New Ideas' (1984) 45 *University of Pittsburgh Law Review* 499, 503.

[203] Montgomery (n 69).

[204] Ibid.

[205] Ibid 6.

[206] Land Reform (Scotland) Act 2003.

but pragmatic answer to spatial unevenness. Other jurists propose analogous abstractions, places that are an entrance to community, not an exit,[207] or progressive concepts such as human flourishing.[208] Of course, public wealth is also an ideal, a utopian-like yearning for a spatial life lived well. Like Massey's places, redolent in their 'progressive sense of place',[209] or Cooper's 'everyday utopias', sites of constant place-making, such as free-thinking schools or queer bath-houses, 'conceptually potent, innovative sites [which] can revitalize progressive and radical politics through their ability to put everyday concepts (such as property . . .) into practice in counter-normative ways'.[210] Public wealth also occasionally enters political discourse, such as the UK Labour Party's 2019 election manifesto, which incorporated public wealth, 'private sufficiency and public luxury'[211] into its policy calculations. The antecedence of these concepts can be located in claims for the promotion of the *commonweal* and 'communal luxury'[212] as discussed in Chapters 3 and 4. Public (or common) wealth is ultimately a crude spatial trade-off: '[t]he expansion of public wealth in land creates more space for everyone, while the expansion of private wealth in land reduces the space available for others.'[213] The public more, it seems, is the merrier place. The private more is conversely not.

The places of the lawful forest are also places under threat or siege. They may be on the urban interface, for example inner suburbs threatened by gentrification.[214] Or they could be places where ancient relational practices face an uphill struggle for survival. In the 'fabled honey forests' of north-eastern Turkey, the ancient beekeeping practices of the Hemshin ethnic minority (where black hives are stored in hornbeam trees in the mountains of Rize) is fast disappearing, under the dual threats of tourism development and deforestation.[215] The Hemshin's practices rely on the use of the hornbeam tree as a

[207] Eduardo M Peñalver, 'Property as Entrance' (2005) 91(8) *Virginia Law Review* 1889.

[208] Alexander et al. (n 86).

[209] Doreen Massey, 'The Patterns of Landownership and Its Implications for Policy' (1980) 6 *Built Environment* 263.

[210] Davina Cooper, *Everyday Utopias: The Conceptual Life of Promising Spaces* (Duke University Press, 2013) 11 (*Everyday Utopias*).

[211] George Monbiot (ed.), Robin Grey, Tom Kenny, Laurie Macfarlane, Anna Powell-Smith, Guy Shrubsole and Beth Stratford, *Land for the Many: Changing the Way Our Fundamental Asset Is Used, Owned and Governed* (Labour Party, 2019) 12.

[212] See analogously Chapter 4 and the discussion of the concept of 'communal luxury' in the Paris Commune.

[213] Monbiot et al. (n 211).

[214] See, e.g., Rebecca Solnit, *Hollow City* (Verso, 2000); Blomley, *Unsettling the City* (n 31).

[215] Daniel Maher, 'Have a Look at the Fabled Honey Forest', *The New York Times* (16 November 2020), available at <https://www.nytimes.com/2020/11/16/travel/turkey-honey-forest-bees.html?searchResultPosition=1>.

year-long hive, 'high above the forest floor [and] out of the reach of any sweet-toothed bears'. Like lawful forests everywhere, the experiences of the honey forests are familiar – and similar. 'Here we find a commonality between the Hemshin people and their bees: Both struggle against homogenization – one for the survival of its culture, and the other just for survival.'[216]

Last, the lawful forest is a place of foment, of struggle, radical politics and a pushing back against enclosure. These are the 'throwntogether here-and-now' moments carved from Keenan's 'subversively shaped spaces',[217] rich in a volatility that sparks conflagration, or else places where spatial injustices remain simmering across generations. Hayes's countryside path is an example of the latter (as well as a microcosm of the macro-English countryside), inter-rupted by the enclosures of Basildon Park and no doubt the source of daily grievance. Access to the river is long gone, but the communitarian memory (and its residual anger) has not faded.

As Blomley argues in his study of the politics of property in Vancouver's Downtown Eastside,[218] and as previously discussed, critical space and place is not politically neutral, but 'continually in a state of contestatory becoming. The landscape is not a backdrop . . . but is itself created through that contest, serving in turn to become a vital symbolic and practical component in future contestations.'[219] Lucy Finchett-Maddock identifies this contest as an incident of propertied injustice – what she calls an enduring pattern of 'proprietorial resistance'.[220] Hartog attributes this to the lingering effects of custom, one of the few political powers of the poor and oppressed.[221] In England, a flurry of recent literature focuses on the political and power imbalances of land and its ownership. The politics of place and property lies at the heart of Brett Christophers's *The New Enclosure*, Guy Shrubsole's *Who Owns England*, and Nick Hayes's *The Book of Trespass*, which collectively study the implications of private hegemony and public impoverishment. These narratives may be the placed tales of today's England, but they are a continuation of an old politics of rural enclosure and spatial hegemony.[222]

---

[216] Ibid.

[217] Keenan, *Subversive Property* (n 5).

[218] Blomley, *Unsettling the City* (n 31).

[219] Ibid 53–4.

[220] Lucy Finchett-Maddock, *Protest, Property and the Commons: Performances of Law and Resistance* (Routledge, 2016).

[221] Hartog (n 37) 899.

[222] Blomley, 'Enclosure' (n 17); Briony McDonagh and Carl J Griffin, 'Occupy! Historical Geographies of Property, Protest and the Commons, 1500–1850' (2016) 53 *Journal of Historical Geography* 1.

Finally, the placed politics of the lawful forest does not simply project backwards, but also imagines forwards. Utopian dreams and utopian theories foreshadow a better if not perfect world. Cooper's 'everyday utopias' illustrate this, how these grounded sites 'support new conceptual lines that lead to different forms of imagining',[223] rich and fruitful performances of transformative, prefigurative politics. Michel Foucault may dismiss utopias as 'fundamentally unreal spaces', yet his counter-vision of the heterotopia is analogous to Cooper's everyday utopia, 'real places' of 'heterogenous sites and relations . . . that take quite varied forms and change over time, as "history unfolds" in its adherent spatiality'.[224] These are the diverse places of the lawful forest; visual, vernacular, political, and often under siege, the 'here-and-now throwntogetherness' that is unique to context, yet faithful to its recurring continuities.

The Phenomenal Tree

This chapter's theoretical excursus ends with the phenomenal tree. The iconoclast English oak, the mighty Australian river red gum, or the towering redwoods of the US Pacific Northwest represent three well-known examples of the genre. Phenomenal trees transcend the material – and confound the secular – with their numinous qualities. Ancient giants that evoke awe and reverence, they underscore the brevity of human life. In Richard Powers's *The Overstory*, two young travellers, on their way to protests on the Lost Coast during the so-called 'timber wars' of the 1980s and 1990s, pass through a grove of giants in northern California:

> The redwoods knock all words out of them. Nick drives in silence. Even the young trunks are like angels. And when, after a few miles, they pass a monster, sprouting a first upward-swooping branch forty feet in the air, as thick as most eastern trees, he knows the word *tree* must grow up, get *real*. It's not just the size that throws him, or not *just* the size. It's the grooved Doric perfection of the red-brown columns, shooting upward from the shoulder-high ferns and moss-swarmed floor – straight up with no taper, like a russet, leathery apotheosis. And when the columns do start to crown, it happens so high, so removed from the pillar's base, that it might as well be a second world up there, up nearer eternity.[225]

Whether we are moved to eloquence, or like Powers's protagonists, rendered speechless by their wonder, phenomenal trees are undoubtedly 'overloaded

---

[223] Cooper, *Everyday Utopias* (n 210) 14. Davies likewise recognises that utopian theory in 'reimagining the world and prefiguring the future', Davies, 'Prefiguring Property' (n 6) 38.
[224] Soja (n 101) 17.
[225] Richard Powers, *The Overstory* (Vintage, 2018) 263–4.

with symbolism'.[226] What does the phenomenal tree signify for the lawful forest? Indeed, why close this chapter with the singular tree, when its constant refrain has been the opposite, not to lose sight of the forest for its trees?

We argue the phenomenal tree is a good place to end for several sound (theoretical) reasons. First, is the phenomenal tree's powerful ability to cut through and 'speak' to us, to disrupt the spatial dynamics of the nature/ culture, subject/object divides, and to recognise its own distinctive 'social identity'. Davies observes that this quality disrupts Hegelian notions that project the human self onto objects of the external world:

> The object is 'external absolutely' and has no ends of its own, no rights, no freedom, no self-consciousness: it is this character of pure externality which, for Hegel, makes every object available for appropriation. And yet, contra Hegel, [the] phenomenal tree is *there*. It *is*.[227]

In its magnificent materiality, the phenomenal tree is unavoidable, it is *there*, 'a [subject-ified] actor in human space and human society'.[228] The phenomenal tree also reminds us of the law's inadequacies in representing it.[229] And, critically, its brooding presence engages and communicates with us. Powers remarks how 'the redwoods do strange things. They hum. They radiate arcs of force. Their burls spill out in enchanted spaces.'[230] Our narrow envisioning of spatial relations leaves no room for engaging with the phenomenal, the non-commodity, the sublime. Yet to view spatial relations differently opens up space for the *other*. Paradoxically, the phenomenal tree 'speaks' in ways that 'ordinary' trees do not, including speaking for the forest as a whole. As Davies surmises, 'its very solidity and thing-ness resists appropriation and representation and promises an internality or inscrutability that, however inaccessible, appears to address us'.[231]

Second, the phenomenal tree is significant because it sits on the cusp of the physical and metaphysical. Or to put it slightly differently, the phenomenal tree has its roots in both worlds, the liminal crossover zones of our lawful forest. Hence, an extant grove of 600-year-old redwoods can be at once 'trunks running upward out of sight' while also being 'the pillars of a russet

---

[226] Davies, 'Consciousness of Trees' (n 9) 219.

[227] Ibid 223.

[228] Ibid 229.

[229] For example, the common law's sparing classification of certain 'timber trees' saved from doctrinal waste, or ad hoc planning rules that notionally protect 'significant trees' – like the doomed river red gum enclosed in Davies's shopping mall atrium.

[230] Powers (n 225) 317.

[231] Davies, 'Consciousness of Trees' (n 9) 222.

cathedral nave'.[232] In this bifurcated way, it acts as a portal into two worlds, a way of seeing the material lawful forest and its metaphysical meanings.[233] In its 'very solidity and thing-ness' the phenomenal tree grounds complex ethereality to the forest floor.

Last, the phenomenal tree is a modern proxy for loss; symbolic of the vast ancient forests felled. Powers says that he was moved to write *The Overstory* by the sheer scale of forest loss. He describes walking in the Santa Cruz Mountains, east of Stanford, and marvelling at the wood's 'wonderful' capacity to recover from its clear felling a century or so earlier to build the city of San Francisco. Yet, over time, he realises that the forest cannot truly recover. Powers tells of coming across a giant remnant tree, by chance overlooked by loggers, and the life-changing implications of this encounter:

> It was the width of a house, the length of a football field, and as old as Jesus or Caesar. Compared to the trees that had so impressed me, it was like Jupiter is to the Earth. I began to imagine what they must have looked like, those forests that would not return for centuries, if ever. It seemed to me that we had been at war for a long time, trees and people, and I wondered if it might be possible for things ever to go any other way. Within a few months, I quit my job at Stanford and devoted myself full-time to writing *The Overstory*.[234]

In England, authors likewise note the survival of great trees amidst the whole-sale destruction of surrounding ancient forests.[235] Derek Niemann describes 'odd little wooded shards that were once part of greater wholes'.[236] Or, poignantly, like Kett's Oak of the Reformation, the 'single grand trees that once marked a woodland boundary' that now sit alone on roadsides, or 'little pockets of woodland, barely copse-sized, "*shadow woods*" wedged among houses and schools'.[237]

---

[232] Powers (n 225) 318.

[233] There is a giant redwood trunk decomposing in Muir Woods, north of San Francisco, said to have fallen on the death of President Franklin D Roosevelt in 1944. In imagining a better world, an early gathering of the United Nations met at the site in the immediate post-war period.

[234] Amy Brady, 'Richard Powers: Writing *The Overstory* Quite Literally Changed My Life', *Chicago Review of Books* (18 April 2018), available at <https://chireviewofbooks.com/2018/04/18/overstory-richard-powers-interview/>.

[235] Peter Fiennes, *Oak and Ash and Thorn: The Ancient Woods and New Forests of Britain* (Oneworld, 2017).

[236] Derek Niemann, *A Tale of Trees: The Battle to Save Britain's Ancient Woodland* (Short Books, 2016) 70.

[237] Ibid 97 (emphasis added).

Phenomenal trees are sentinels of forested pasts, sometimes the last tree standing in their shadow woods. But their fate is also inextricably tied with the planet's survivability, and our precarious futures. In what must be the ultimate dispossession by enclosure, the insatiable end-times appropriation of the earth's commons has pushed us to the brink, the abyss of ecological collapse. Climate change is no longer coming, it is here, or in the words of Sophie Cunningham, 'the unraveling, [i]t's begun'.[238] As this nightmare unfolds, Cunningham clings to trees as she falls off the metaphoric cliff: 'The centre is failing, and we're left to maintain traction in the chaos. Or perhaps the situation is this: we're falling off the cliff in slow motion. Me, I grab at the trees I see on the way down in an effort to break the fall.'[239]

This fall seems somehow more obscene when it takes with it the phenomenal tree – like the former giant now memorialised as a stump big enough for a 'floor where two dozen people danced a cotillion',[240] or the vandalised redwood that became a drive-through tourist attraction. This fall is of course increasingly precipitous, more urgent in its increasing velocity. Climate change in England means that oaks can no longer self-seed in the wild.[241] And summer-long palls of smoke in the US Pacific Northwest[242] are the product of mega-wildfires that in their unprecedented intensity existentially threaten what is left of the giant sequoias and redwoods,[243] forests said to 'store more carbon per acre than any other forest system in the world, by a long shot'.[244] The summer wildfires of 2020 and 2021 posed a threat like never before for northern California's redwoods, once thought immune from firestorms by the (once predominant) climate of 'misty forests, cool ocean breezes and midsummer fogs'.[245] Headlines in *The New York Times* could not be clearer for the fate of these 'charismatic megaflora', and the inextricable connections that bind us all. 'These Trees Are in For The Fight of Their Lives.

---

[238] Sophie Cunningham, *City of Trees: Essays on Life, Death, & The Need for a Forest* (Text Publishing, 2019) 90.
[239] Ibid 13.
[240] Powers (n 225) 249.
[241] Fiennes (n 235).
[242] Jessie Kindig, 'Defensible Space', *Boston Review* (22 October 2018), available at <http://bostonreview.net/science-nature/jessie-kindig-wildfires-western-myth>.
[243] See, e.g., Cunningham (n 238) 146–7.
[244] John Branch, 'They're Among the World's Oldest Living Things. The Climate Crisis is Killing Them', *New York Times* (9 December 2020), available at <https://www.nytimes.com/interactive/2020/12/09/climate/redwood-sequoia-tree-fire.html?action=click&module=Top%20Stories&pgtype=Homepage>.
[245] Ibid.

The Climate Crisis is Killing Them'.[246] In September 2021, firefighters took the highly unusual step of wrapping a number of California's iconic monarch sequoia trees in fire-resistant blankets to protect them from incoming wild-fire.[247] Amongst these protected trees was 'General Sherman', believed by the US National Park Service to be the largest tree in the world – it is 1,487 cubic metres, 84 metres tall, and has a circumference of 31 metres at ground level.[248]

Powers's *The Overstory* concludes with a tale of two protestors who spent weeks aloft in the forest canopy, strapped onto an unwieldy platform in a redwood called 'Mimas'. Mimas was phenomenal, 'wider across than [Nick Hoel's] great-great-great-grandfather's old farmhouse'. Its setting was a con-fluence of the worldly and other-worldly:

> Here, as sundown blankets them, the feel is primeval, darshan, a face-to-face intro to divinity. The tree runs straight up like a chimney butte and neglects to stop. From underneath it could be Yggdrasil, the World Tree, with its roots in the underworld and crown in the world above.[249]

Yet for all its physical thing-ness and metaphysical awe, Mimas, the subjecti-fied phenomenal tree, is ultimately felled, destroyed by a linear mindset of 'progress' and its insatiable urge to consume and enclose. Mimas's fate is the brooding threnody of our times, a lament for its loss, and the world that is lost with it.

Conclusion

The lawful forest is simultaneously ubiquitous and invisible, a spatial para-dox. Its invisibility is (mostly) down to deliberate structural design, a landed politics that for centuries has privileged certain spatial relationships as prop-ertied, and others not. By contrast, the lawful forest's ubiquity becomes obvious only once these machinations are seen for their partisan ways. In large part, this *revealing* has been the work of this chapter. Since away from the lawful forest, on the cleared private plains, fences and walls have come to normalise spatial relationships, while lines of sequestration and exclusion harden the parameters of spatial injustice. And as the fate of the phenomenal

---

[246] Ibid.
[247] Gabrielle Canon, 'World's Largest Tree Wrapped in Fire-Resistant Blanket as California Blaze Creeps Closer', *The Guardian* (18 September 2021), available at <https://www.thegu ardian.com/us-news/2021/sep/17/worlds-largest-tree-wrapped-in-fire-resistant-blanket-as -california-blaze-creeps-closer>.
[248] Ibid.
[249] Powers (n 225) 325.

tree portends, this paradigm's end-times is enclosing fast on our earthly commons.

Flip the perspective, however, and a very different spatial relationship emerges into plain sight. One of 'wide-open synergies', of ancient relational practices, and a common-wealth that expands exponentially in proportion to the public acreage preserved. Seeing the danger of 'fences multiplying across the land' – as Henry David Thoreau warns in his 1862 *The Atlantic* essay on 'Walking' – is to recognise the risks of unchecked enclosure, the implications of a spatial monopolisation that this book traces from the swine pastures of pre-Norman England to the urban forests of the twenty-first century. This is our critical history of (private) property, (the ground-up resistance) of protest, and the imperatives of spatial and environmental justice.

In theorising the lawful forest, the milieu of critical space and place is where these *other* landed relationships subsist – and consistently flourish. These are the 'simultaneity of stories-so-far', which from time to time, and across space and place, recur. Enlivened by the many, not the few, they 'occur across the grain' – lateral traversals characterised by shared common features, the likes of communitarian sociability, a ground-up legitimacy, the crucible of context, and so on. This intersection is the theoretical premise of the lawful forest, a space and place where the imaginings of the *other* are periodically performed and enacted. Their stories run counter to dominant narratives. Their lessons we are exhorted to ignore or overlook, lest the hidden lawful forest be seen for what it is amidst its individual trees.

2

The Ancient Forest

ANCIENT ELM

In writing of eleventh-century French law and society, jurist and historian Geoffroi Flach observes, '[i]l est peut-être vrai de dire que la liberté est sortie du fond des bois.'[1] Flach's reflection on ancient verities, of freedoms emanating from 'the bottom of the woods', speaks to a truth at the heart of the lawful forest. As the Introduction and Chapter 1 outline, our tale explores the intersection of ancient custom and the public square, and the social and political freedoms that populate this subliminal, contested, inherently potent space. Like Flach's aphorism, our narrative, and this space, is never far from the forest – whether long-felled Norman woods, or the remnant forest that is twenty-first-century London.[2] In this chapter, we write of a vast English acreage of Oak and Ash and Thorn,[3] a forest long gone, scarcely imaginable. Felled first to the till, then to the town,[4] this forest nonetheless persists in its canopied memories. In reaching back a millennium and more, this chapter's aim is to reconstruct (and reconceptualise) what lies at the bottom of these long-disappeared woods.

If 'the past is a foreign country',[5] this ancient place seems an alien world. Yet, at some levels, it is not. While separated by the vast gulf of time, we

---

[1] 'It is perhaps true to say that freedoms come from the bottom of the woods', G J Flach, *Les origines de l'ancienne France*, cited in Reginald Lennard, *Rural England 1086–1135: A Study of Social and Agrarian Conditions* (Oxford University Press, 1959) 255.

[2] In a curious coincidence, Detroit's early founders included French-speaking *coureurs des bois* or 'runners from the woods', now a sub-culture in Michigan identifying as French Metis or the Muskrat French. These same fur-traders were cited by Harold Demsetz to justify the pre-eminence of private property rights.

[3] Peter Fiennes, *Oak and Ash and Thorn: The Ancient Woods and New Forests of Britain* (Oneworld, 2018).

[4] Much of this loss is modern, in the years from 1945 to 1985; 'Britain lost more ancient woodland in 40 years than in the previous 400', Derek Niemann, *A Tale of Trees: The Battle to Save Britain's Ancient Woodland* (Short Books, 2016) 8. Conifer plantations, changes in agricultural practices, motorways and urban expansion all played their part.

[5] This saying also takes another form: 'the past is *another* country'.

share a common legal landscape, a *locus* where ancient memories were once enacted, and now linger in the shaded margins – *ghostly* in their apparition. As Sarah Keenan argues:

> to acknowledge and study ghostly matters is important in recognizing the complex ways that power operates ... from state institutions and inescapable meta social structures such as racism and capitalism, through [to] countless, seemingly innocuous everyday things, practices and understandings.[6]

The customs at the bottom of Flach's woods comprise deeply held, commonly understood memories of ancient 'everyday practices and understandings', hauntings that underline 'how that which appears to be not there is often a seething presence'.[7] This is why, to better understand this *seething presence*, and its significance to our modern-day lawful forest, we reach back to the distant past – to paint a picture of the customary history that this collective memory draws on; the forest customs that pre-dated (and survived) the Norman Invasion, and those same ancient customs that were reasserted under the Forest Charter of 1217.

The Forest Charter itself is a remarkable instrument, and not merely because of its antiquity. Its significance lies, as intimated, in the rights that it recorded and restored – providing us with both evidence of this history and a contemporary signpost of a still seething, haunting presence. Of arguably equal significance is the Charter's obscurity, its near-invisibility across the centuries. As Lucy Finchett-Maddock posits, why did its better-known counterpart, the Magna Carta, successfully rid itself of its 'communal twin'?[8] As this chapter argues, the Forest Charter has been largely ignored because it tells a tale at odds with property's dominant discourse. The Charter's invocation of common rights and the 'liberties and free customs of the forest' runs counter to property's universalising narratives: those of private rights, enclosure and alienation. The Charter's obscurity also mirrors an equally obscure (and inconvenient) history of communal property rights and customs, a parallel narrative that stubbornly endures, much like the Charter.

In writing of the lawful customs of Anglo-Saxon England, the subsequent Norman forest laws, and the little-regarded Forest Charter of

---

[6] Sarah Keenan, *Subversive Property: Law and the Production of Spaces of Belonging* (Routledge, 2015) 167.

[7] Ibid.

[8] Lucy Finchett-Maddock, *Protest, Property and the Commons: Performances of Law and Resistance* (Routledge, 2016).

1217, this chapter takes us back, improbably, to the very doorstep of 'time immemorial'.[9] The legal record from this distant past is scant and patchy. Historians describe Anglo-Saxon laws as 'a mass of legal custom', a decentralised, *heterogeneous* jurisprudence faithful to its villaged context. From what little remains, we can only surmise the lie of this customary terrain. Indeed, in staring at the dark spaces between the law's few sparse lines, we must trust to our intuition, conceptualising ancient custom in ways analogous to how we might imagine (and savour) the first millennium's pure forested air.

We begin in the year 825, a sufficiently ancient date to start the lawful forest's ancient tale. It is an arbitrary place to commence, as random as the record that survives. In this year, a case note (for want of a better term) evidences a dispute between the king of Mercia, and a bishop and his community. The contest is over the resources of the community's woods, its catalyst the king's urge to encroach into the wood-pastures with his herd of swine. In its quirky way, this dispute foreshadows the millennium ahead, of recurring clashes between the forces of private enclosure and a public counter-resistance.[10] From the king of Mercia to the eve of the duke of Normandy, similar micro-fragments echo what remains of the pre-Norman customary legal record.

Having painted a picture of the pre-Norman legal landscape, this chapter moves on to the Norman Invasion of 1066 – a timeless tale of the conquered and the conqueror, a seminal moment of disruption when a new legal regime supposedly displaced the old. Conventional rhetoric paints an image of the *tabula rasa*, a legal landscape emptied of its propertied past, a freshly blank canvas over which foreign feudal tenures were imposed. While feudalism is a property law truth, elsewhere the Normans preferred political convenience

[9] 'Legal memory' at common law is limited to 6 July 1189, the accession of Richard I. Events before that date are said to be beyond time immemorial. The term 'time immemorial' has likewise been used to describe the duration of Indigenous laws within the context of their recognition in Australian common law. For example, in *Milirrpum v Nabalco Pty Ltd* (1971) 17 FLR 141 (see Chapter 5), the plaintiffs claimed that they had 'occupied the subject land from time immemorial as of right' (150). The court observed that the phrase was used 'perhaps somewhat unhappily; at any rate the technical connotations of that phrase in English law had no relevance' given that Indigenous occupation described a 'period going back into the indefinite past' (152).

[10] 'Karl Polanyi in his 1944 book *The Great Transformation* identified that human societies tend to react forcefully to the privatization of land [and] to its treatment as a commodity to be bought and sold in markets . . . He described this type of forceful reaction as a "counter-movement"', Brett Christophers, *The New Enclosure: The Appropriation of Public Land in Neoliberal Britain* (Verso, 2018) 324.

or legal pragmatism, in so doing leaving vast swathes of custom dormant on the forest floor. Indeed, where the conqueror's laws were unpopular, like the egregious forest laws – that transformed large portions of the English landmass into royal game parks – ancient custom resisted with potent effect. The Forest Charter of 1217 is a powerful reminder of what lay beneath, its restoration of the rights and privileges of 'the forest's good and lawful men'[11] providing – by inference – yet other missing pieces in the incomplete customary jigsaw.

At its heart, this chapter explores the vexed tensions between custom and the law[12] – by reaching back to a time when custom *was* the law. Its intent is not to yield a sweeping account of earliest English legal history. This is done elsewhere, in greater detail, and with faithful deference to sovereigns and their chronology.[13] Rather, this chapter's objective is to reveal a fundament of the modern common law: that beneath its veneers, its uppermost layers of acts, regulations and black-letter certainty, the law has an ancient customary past – and present. Our lawful forest originates here, deep beneath the forest's leafy floor, a *locus* where customary rights and communal freedoms find fertile soil, a rarely visited yet profoundly rich place 'au fond des bois'.

The Anglo-Saxon Lawful Forest

It is said that '[w]hen William the Conqueror seized the English Crown, he became the ruler of an ancient realm. England was already an old and settled country.'[14] Likewise, it had old and settled laws. The laws were not those of a unitary state, but rather a heterogeneous sum of loosely connected parts. The surviving records of *written* Anglo-Saxon laws are small in number, ad hoc in content, and 'patently unrepresentative' in their nature.[15] There is a

---

[11] The historical record appears to indicate that it was mostly *men* whose rights and privileges were recognised and restored by the Forest Charter of 1217. Unearthing what belonged to the women of this time, both in practice and at law, is a hugely worthy project, but one that is beyond the scope of this book.

[12] See the discussion in the Introduction, and Edward Coke's 'project' to discredit and delegitimise custom as a basis of the common law.

[13] See, e.g., Robin Fleming, *Domesday Book and the Law: Society and Legal Custom in Early Medieval England* (Cambridge University Press, 1998); Patrick Wormald, *The Making of English Law* (Wiley-Blackwell, 2001).

[14] Lennard (n 1) 1.

[15] Tom Lambert, *Law and Order in Anglo-Saxon England* (Oxford University Press, 2017) 13. There are three main repositories: in Rochester Cathedral, Corpus Christi College, Cambridge, and the British Library, Richard Gameson, *The Manuscripts of Early Norman England (c. 1066–1130)* (Oxford University Press, 1999).

disproportionate concentration of edicts surviving from the Kentish or West
Saxon (Wessex) kingdoms. Nonetheless, these 'vernacular documents possess
special value . . . by throwing light upon the nature . . . of the unwritten
customary law'.[16] What remains are narrow windows, casting tiny slivers of
light, into a long-past customary forest.

Early twentieth-century English historian Dorothy Whitelock describes
Anglo-Saxon laws as:

> a mass of customary law with various local differences, but supplemented
> and modified by the statutes of successive kings; . . . from seventh century
> Kings of Kent, from Ine of Wessex, from Alfred, and from most of his
> successors up to Cnut.[17]

Custom's 'various local differences' arose from a panoply of diverse jurisdictions: small kingdoms,[18] the domains of powerful aristocrats, the privileges
of towns, the conventions of churches or secular guilds, and above all, the
unwritten rules of 'communities of free men'.[19] The Anglo-Saxon legal landscape was thus a *sui generis* customary polyglot, a 'coherent, stable and enduring legal order neither state-like nor stateless', an order 'best understood on
its own terms'.[20] Legal geographer Nicholas Blomley likewise describes the
pre-Norman jural map as:

> decentered, pluralistic and often extra-statal . . . Pre-Norman rulers were
> content to leave the shire and hundred courts as the main locus of justice.
> These jurisdictions were supplemented by other multiple and overlapping
> 'legal spaces' that appear foreign to the modern state. Many of these were in
> the form of private jurisdictions . . . Manorial courts were also widespread,
> operating according to their own customs and procedures. The metaphor
> of the 'forest of spires and towers' seems apposite, the courts of justice were
> local and resort to any central tribunal was rare.[21]

---

[16] A J Robertson (ed. and trans.) *Anglo-Saxon Charters* (Cambridge University Press, 1939).

[17] Dorothy Whitelock, 'The Anglo-Saxon Achievement', in D Whitelock, D C Douglas,
C H Lemmon and F Barlow (eds) *The Norman Conquest: Its Setting and Impact* (Eyre and
Spottiswoode, 1966) 26.

[18] Many of which were absorbed over time into larger, more successful kingdoms, such as the
kings of West Saxon.

[19] Lambert (n 15) 299–303, 349.

[20] Ibid 156. Levi Roach concurs, noting that legal historian Patrick Wormald (writing in
1977) 'realized more clearly than any before him how . . . different later Anglo-Saxon legal
culture was from its modern counterpart', Levi Roach, 'Law Codes and Legal Norms in
Later Anglo-Saxon England' (2013) 86 *Historical Research* 465.

[21] Nicholas Blomley, *Law, Space, and the Geographies of Power* (Guildford Press, 1994) 77–8.

Sovereign interventions into this diverse patchwork were few – misnamed 'codes'[22] that were in substance clarifications or modifications of an otherwise near-universal custom.[23] As a matter of practice, there was little need to reduce to writing laws that 'every freeman learnt by oral tradition'.[24] Kings were similarly reticent to legislate because of the prevailing paradigm: 'The ancient customs and laws of the people were not regarded as alterable at the whim of a temporary human ruler. We have seen that Alfred said, "I durst not set down much of my own."'[25]

Alfred's reluctance to overlay unwritten custom with written law reflects the privileging of the oral tradition in Anglo-Saxon jurisprudence. Tom Lambert explores how oral custom was maintained over five centuries. He proffers two scenarios that explain its endurance, first the role of wise 'old men [who] would have known what the law was by virtue of their years of legal experience'.[26] Such elders were sought by kings or councils for their advice. Whitelock (in part) corroborates this theory, citing the example of 'the bringing of the aged bishop Ethelric of Selesy to the plea held on Pinnendon Heath in 1075 or 1076' to answer Norman queries as to the content of certain Anglo-Saxon laws.[27] Lambert's alternate explanation goes to the memorisable nature of *peawas* (or custom), especially in Kentish law, which featured linguistic patterns that facilitated eas(ier) learning by rote.[28]

If written laws merely clarified custom, then the dearth of forest references in Anglo-Saxon codes problematises how we can conceive of the unwritten customs of the forest. Apart from occasional references to the forests of the Weald and Selwood, or the notation of 'leys' or forest clearings, 'little is said in the records of the great forests themselves'.[29] As Peter Blair observes, 'nothing comparable with the forest laws of later times is known from the Anglo-Saxon period'.[30] What endures is a mere handful of references, a 'fragment of the customary law'.[31] In the West Saxon Laws of Ine

---

[22] 'Anglo-Saxon legal codes were not so much concerned with the promulgation of new law as with the codification of existing custom', Peter Blair, *An Introduction to Anglo-Saxon England* (Cambridge University Press, 1956) 235.

[23] W J V Windeyer, *Lectures on Legal History* (Law Book Co., 1957).

[24] Ibid 6.

[25] Ibid 7.

[26] Lambert (n 15) 31.

[27] Whitelock (n 17) 27.

[28] Lambert (n 15) 69–70.

[29] Blair (n 22) 250.

[30] Ibid.

[31] C Stephenson and F G Marcham (eds) *Sources of English Constitutional History: A Selection of Documents from AD 600 to the Present* (Harper and Bros., 1937) 1.

(dating from 688 to 695), there are four references to the woods. Chapter 20 deems 'a man from afar, or a stranger' who 'travels through the woods off the highway' to be 'a thief unless he calls out or blows a horn'.[32] Chapter 43 indicts 'anyone who burns down a tree in the wood' liable to fine 'for fire is a thief'.[33] However, where the felling is productive, not wanton, the feller must pay for 'three trees at 30 shillings each' but 'need not pay for more of them however many they were, for the axe is an informer not a thief'.[34] Where the woods could be used gainfully as swine pastures, the punishment was increased, chapter 44 declaring that 'if anyone however cuts down a tree under which 30 swine could stand, and it becomes known, he is to pay 60 shillings'.[35] Historical records consistently refer to the keeping of herds of swine, such that 'extensive pig-farming was a normal feature of districts in which there was much glandiferous woodland'.[36]

In the later Laws of Alfred (871–99) there is only a single reference to woods. Clause 12 punished a transgressor thus: where 'a man burns or fells the wood of another, without permission, he is to pay for each large tree with five shillings, and afterwards for each, no matter how many there are, with fivepence, and 30 shillings as a fine'.[37] These fragments of written forest law speak to the woods as a valuable natural resource, a source of timber for shelter or warmth, and in their wood-pastures, a forage resource for grazing pigs. They also speak to community and communitarian relationships, and how one man's theft of trees poisoned the communal well through deceit and mistrust. Theft was said to be a particularly hated crime in Anglo-Saxon legal culture, a policy based on 'the high value placed on communal solidarity'.[38] Of course, these 'woods of another' were not privately owned woods, since it would take another millennium for 'a clear and unqualified definition of private property' to appear in 'the works of any legal writer'.[39] Rather, these

---

[32] Failure to so signal his intent renders this stranger liable to be 'slain or redeemed by payment of his wergild', ibid 8.

[33] David Douglas, *English Historical Documents: Vol. I, c.500–1042* (Eyre & Spottiswoode, 2nd ed., 1979) 403.

[34] Ibid 403–4.

[35] Ibid 404.

[36] Lennard (n 1) 255. Glandiferous woods feature trees that are nut bearing.

[37] Douglas (n 33) 411.

[38] Lambert notes that 'hatred of theft was a deeply rooted aspect of Anglo-Saxon legal culture ... that may well reflect the high value they placed on communal solidarity', Lambert (n 15) 99–100. Theft poisoned the fabric of trust and relationships between neighbours that weakened community, ibid 100–7.

[39] Nicholas Blomley, 'Making Private Property: Enclosure, Common Right and the Work of Hedges' (2007) 18 *Rural History* 1, 4.

woods were community resources, over which community members enjoyed privileged use rights. These woodlands were forest commons.

Herds of swine and entitlements to communal woods were at the heart of the so-called 'Lawsuit about Wood-Pasture' of 825.[40] In this 'very note-worthy suit' about a wood-pasture at Sinton (today a village known as *Leigh* Sinton, or the clearing at Sinton), a dispute arose over the extent to which King Beornwulf's herds of swine could encroach into the wood-pastures of the Sinton community. The pleadings indicated that 'the (king's) reeves in charge of the swineherds wished to extend the pasture further, and take in more of the wood than the ancient rights permitted'.[41] The action, brought by the bishop and the community's 'advisers', argued that the extent of ancient custom (dating to King Ethelbald) permitted only 300 swine to graze in the wood-pasture. And of those 300, two-thirds were reserved to the bishop and the community. An oath as to the content of that ancient custom was furnished to the 'bishop's see at Worcester' thirty days after the original council meeting. The king's reeve in charge of the Sinton herd of swine, a person known as 'Hama', was present at the later declaration of the oath, and according to the translation, he 'watched and observed the oath . . . but did not challenge it'.[42]

The community's wood-pasture at Sinton proved to be no ninth-century tragedy of the commons. Customary rules ensured that the number of grazing pigs did not exceed a sustainable herd, in terms indicative of later common rights of pannage.[43] Moreover, in an early affirmation of the rule of law, ancient custom was sufficiently robust to resist royal encroachment. This is indeed a 'very noteworthy suit', in what it infers about the ancient customs of the Sinton woods. First, is the priority accorded to the pasturage resource, a privileging that overcame the king's plans to extend his private holdings. Second, is the notion of community 'ownership' of the woods, a title suf-ficient to give standing to institute and successfully prosecute the lawsuit. Elsewhere, other surviving fragments speak to community customary owner-ship, a lease of land to Edmund Etheling by the Community at Sherbourne,[44] or an agreement between the community at Worcester and Fulder, which granted the latter a three-year usufruct, or use-right.[45]

---

[40] Robertson (n 16) 9.

[41] Ibid.

[42] Ibid.

[43] *Pannage* refers to the putting of swine into the woods to feed.

[44] Robertson (n 16) 147.

[45] Ibid 155. At the end of the three-year term, the estate was to be returned to the community 'with the full equipment supplied him by the community, namely 12 slaves and 2 teams of

Today, such notions seem remarkable. First, is the interpellation of a communitarian ownership that disrupts the liberal binary of the self-maximising individual and the regulatory state.[46] Second, is custom's role in protecting communal resources from over-use, an account at odds with Garrett Hardin's contrived tragedy of the commons. The former reminds us, as seen in Detroit's urban farms, that community commons are complex and contested places, with vexed understandings of insiders and outsiders, belonging and non-belonging, and competing accounts of over-use, under-use and definitions of tragedy. The latter reminds us of the flaws in Hardin's thesis, the inconvenient examples, historic and modern, of sustainable common practices, like the seasonal obligations of 'stinting' that managed stock levels on common pastures.

If, as we contend, the common law has an ancient customary past, one as ancient as the customs of pre-Norman England, then how do we fill in the vast gaps between the threadbarest of written records to depict what lies at the bottom of the Anglo-Saxon woods, the source of Flach's ancient freedoms? Lambert's 2017 book *Law and Order in Anglo-Saxon England* is a contemporary attempt to reconstruct Anglo-Saxon criminal law. Lambert acknowledges the enormity of his task, a millennium or more after the event:

> It is highly difficult to construct narratives of legal practice before the 10th century . . . Even in the later Anglo-Saxon period, if we accept conventional wisdom . . . the best we can hope for is a partial understanding of some late Anglo-Saxon legal culture and practice. The only way to expand our field of vision is to examine how people behaved in a later, better-documented age and project it backwards. Starting at the beginning *and* taking a holistic approach seems an impossible combination.[47]

In a similar vein, to reconstruct a narrative of forest custom requires an interpretative approach that takes a leaf from Lambert's book, one that enhances 'analytical opportunity' and 'interrogates the discourse of the laws productively'.[48] Lambert suggests a number of devices. First, is to view existing fragments of written law as 'much richer sources than conventional wisdom would suggest . . . [to tease] a great deal of value out of them if approached sensitively and imaginatively'. Second, is to recognise that 'Anglo-Saxon legal

---

oxen and 100 sheep and 50 fothers of corn'. The final recitation reads: 'And if anyone breaks this agreement, he shall never be forgiven, but shall be condemned to the torment of hell, and shall dwell there with the devil until Doomsday', ibid.

[46] Gerald Frug, 'The City as a Legal Concept' (1980) 93 *Harvard Law Review* 1057.

[47] Lambert (n 15) 13.

[48] Ibid 14.

discourse was not just inhabited by the aristocrats . . . but by non-elite society more generally, or at least by the legally free'. Third, as noted in the quote above, is to 'examine how people behaved in a later, better-documented age and project it backwards'. Last, is to acknowledge that laws are a much more useful category of evidence where 'their context is viewed in equal weight to the texts themselves'.[49]

The first two approaches reflect their milieu, gleanings taken from the meagre leftovers of the times. Invoking 'imagination' and 'sensitivity' is akin to what Paul Carter calls 'staring at dark writing', the blank spaces between the lines of textual imprint. In writing of another ancient land, the ancient place and laws of Indigenous Australians, Carter says that 'dark writing indicates the swarm of possibilities that had to be left out when the line was taken',[50] a 'dappled', profoundly nuanced seeing of things. Earlier assumptions made about the value of forest resources to community, or how forest theft threatened community cohesion, provide some hint of the swarm of customary possibilities left out when the few legal lines were taken. Yet they cast more shade than dappled light.

Lambert's third device, to take subsequent, better-documented evidence and 'project it backwards', yields a richer harvest of forest custom. One source is the Domesday Book, that 'grand survey of the English kingdom conducted in 1086' that contains, *inter alia*, brief summaries of the customs of several shires.[51] Another is the incidents of later common property rights. And yet others are the contents of Norman-era forest laws, and the reformist Forest Charter of 1217, the subject of separate and later discussion in this chapter. Collectively, these laws showed how people behaved in analogous context, and through retrospective implication, infer how forest custom may have regulated the affairs of free men.

The Domesday Book was the output of a royal commission ordered by William in 1085. This grand survey of the kingdom generated a mass of statistics, a list of facts and figures pertaining to 'grants, livery, mortgages, sales, antecession and the like'.[52] It was a survey of property, or more particularly,

---

[49] Ibid.

[50] Paul Carter, *Dark Writing: Geography, Performance, Design* (University of Hawai'i Press, 2009) 1. In speaking of lines, Carter refers to the lines of the paintings of Western desert artists, studied in the early 1970s: 'To read a Western desert painting was to apprehend the performance of that story. This was the success of the painters – to re-present a haptically documented, ceremonial event in a visual form, to find a language that communicated permanently an ephemeral performance', ibid 123.

[51] Lambert (n 15) 166.

[52] Fleming (n 13) 3.

a survey of property disputes and the ownership anxiety unleashed by the Invasion.[53] Yet, as Robin Fleming argues, the commissioners were arguably more 'concerned with the survey's more anecdotal information' than the 'brutalist prose' or 'mind numbing detail' of its data.[54] Fleming says that legal historians largely overlook that the Domesday Book is 'the most comprehensive, varied and monumental legal text to survive from England before the rise of the Common Law'.[55] In the case of this chapter, it explains the ways in which 'old, familiar legal customs' defended property and reinforced the 'importance of living memory in law and for law's pronounced sociability'.[56] Domesday also reveals wide regional variations in custom. Most importantly, Fleming argues that Domesday bolstered continuity in English land law,[57] an outcome that suited the political imperatives of the ruling clique. What came out of Domesday were the founding conventions of Anglo-Norman land law, a 'hardy and flexible mix' of Norman and (predominantly) English custom. Domesday is replete with references to 'woodlands', 'groves', 'woods' and 'underwoods', and rewarded landholders who '[took] care of forest'.[58] As Fleming notes, Domesday is significant not for its explicit detail of title grievances, but for the narratives underlying it, and the broad continuity it enabled.

If Domesday supplies continuity, later common property rights provide a rear-view mirror of elaborate detail. Often described as the incidents of communitarian custom, these common rights were not public rights, but rather the exclusive entitlements of eligible community members to use and share common resources. They were also rights to exclude outsiders. The enclosure of open wastes, the felling of woods and the draining of fens, all in the name of 'improvement', extinguished a vast diversity of common tenures from the legal landscape. Yet, in ways reminiscent of the resilience of public custom, traditional common tenure survives to the twenty-first century. Indeed, as Carol Rose argues, common property has proved adaptive to modern contexts, re-forming as member rights of urban housing co-operatives, social or

[53] 'Ownership anxiety' is a term coined by Rose to describe one way in which William Blackstone's famous prescription of property operated. Carol Rose, 'Canons of Property Talk, or, Blackstone's Anxiety' (1998) 108 *Yale Law Journal* 601. Anxiety over valid and secure title is an inevitable consequence of invasion, where one group's title holding is supplanted by another's. Inevitably, this is a reality that the state likewise seeks to assuage. In the case of the Norman Invasion, this was a central objective behind the Domesday Book.

[54] Fleming (n 13) 3, 5.

[55] Ibid 5.

[56] Ibid 63.

[57] Ibid 83.

[58] For example, paragraph 563.

sporting clubs, or community gardens or orchards. Beneath the ostensibly legal surface, we begin to see how an ancient lawful past may linger, and haunt.[59]

Use rights such as pasturage were the most prevalent form of rural common tenure.[60] Other common rights included the gathering of wood, turf or gorse for heating,[61] the felling and collection of timber for repair of fences and houses,[62] fishing in common streams,[63] access across the countryside,[64] gleaning,[65] foraging for wild foods such as nuts, mushrooms and berries, the collection of honey and wax from forest hives, passive recreational uses (such as rights of air and exercise), and the festive use of open space.[66] As Jeanette Neeson surmises, 'the diversity of common rights varied according to local *custom* and usage'.[67] Richard Hayman explains that '[i]n practical terms this meant that common rights differed from one parish to another.'[68] Such diversity often took idiosyncratic form; for example, in 1273 in the Forest of Arden, an enquiry found that while pannage was forbidden in the woods,

---

[59] Carol Rose terms these rights 'modern limited common property' – rights that are more ubiquitous than we suppose, Carol Rose, 'The Several Futures of Property: Of Cyberspace and Folk Tales, Emission Trades and Ecosystems' (1998) 83 *Minnesota Law Review* 129–82. See also generally, John Page, *Property Diversity and Its Implications* (Routledge, 2017) 70–92.

[60] *Pasturage* generally referred to 'cattle', being animals needed to plough the land (horses), or animals needed to manure the arable land, such as *cattle proper*, and sheep. Variants to *pasture* include vesture and herbage. Rights of pasture could be limited to specific numbers of cattle through 'stinting', Jeanette Neeson, *Commoners: Common Right, Enclosure and Social Change in England, 1700–1820* (Cambridge University Press, 1993) 114. Alternatively on some lands, rights of pasture could only be exercisable at specified times of the year ('commonable lands'), Navjit Ubhi and Barry Denyer-Green, *Law of Commons and of Town and Village Greens* (Jordan Publishing, 2004) 27–30. Pigs (pannage), geese and goats were separate common rights.

[61] Turbary.

[62] Estovers (including house-bote, cart-bote and hay-bote).

[63] Piscary.

[64] 'To local people almost entirely employed in agriculture . . . there simply was no notion of rights of access onto the common separate from the rights and incidents of rights of common. With the massive urbanisation of the 19th century, and the disengagement of many people from agricultural activities, many commons came to be seen to have separate or additional recreational purposes', Ubhi and Denyer-Green (n 60) 133.

[65] The collecting of the remains of a harvest left in the fields.

[66] Customary activities include cricket, Morris dancing, Maypole festivals and village fairs, L Dudley Hoskins and William Stamp, *The Common Lands of England and Wales* (Collins, 1963) 4.

[67] Neeson (n 60) 313 (emphasis added).

[68] Richard Hayman, *Trees: Woodland and Western Civilization* (Bloomsbury, 2003) 148.

local tenants could still 'gather enough oak and thorn branches between Martinmas and Easter for fencing to last for two years', and otherwise they were unrestricted to 'gather dead wood for domestic heating, baking and brewing'.[69] Meanwhile, in Epping Forest, the common right to graze overrode a landowner's rights of management, such that the landowner could only 'pollard their trees rather than coppice them'.[70]

Commoners derived not only material sustenance from their commons – food, timber and fodder – but also a sense of place, a social capital or 'knowledge'. For example, Peter Linebaugh describes how commoners 'wandered out of their knowledge' when leaving common lands,[71] the concept of 'knowledge' being the very apogee of custom – communitarian norms that are deeply implicit, universally understood, and somehow never forgotten.[72] Likewise, Hayman draws this link between commons and custom, describing commons as the 'plebeian underwood', and common rights 'evolving gradually by customary use so even in the thirteenth century their origin could be described as *time-hallowed*'[73] – imbued with a sense of timelessness. Woodlands were 'busy places' in ancient England, an integral part of rural economies, and fundamentally integrated with rural life itself.[74] Their customary regulation and the protection of 'plebeian' rights to their 'underwood' provide a rich insight into unwritten forest customs and their deeply intricate content – documented from the benefit of (proximate) hindsight.

The significance of context, Lambert's fourth interpretative tool, is likewise insightful in discerning custom's emplacement in landscape. This field of enquiry is now the province of legal geography, a material view of law and society that 'makes the interconnections between law and spatiality . . . especially their reciprocal construction'.[75] Through this prism, the law is seen to be 'located, [it] takes place, is in motion or has some spatial frame

---

[69] Ibid 148–9.

[70] Ibid 149. To 'coppice' a tree involves pruning it near the ground, whereas a 'pollarded' tree is cut close to its head.

[71] Peter Linebaugh, 'Enclosures from the Bottom Up' (2010) 108 *Radical History Review* 11, 19.

[72] For example, in twenty-first-century Australia, specifically in areas surrounding national parks in the NSW Northern Rivers, classes are advertised offering the development of skills to 'forage on public lands', 'utilizing indigenous and non-indigenous flora' and 're-discovering ancient practices', Jacqueline Munro, 'Forage for Bush Food: Expert Says He'll Show You How', *The Northern Star*, 4 March 2019.

[73] Hayman (n 68) 148 (emphasis added).

[74] Ibid 147.

[75] Irus Braverman, Nicholas Blomley, David Delaney and Alexandre Kedar (eds) *The Expanding Spaces of Law: A Timely Legal Geography* (Stanford University Press, 2014) 1.

of reference'.[76] Importantly, 'social spaces, lived spaces and landscapes are inscribed with legal significance' such that they cease being 'inert sites of law but inextricably implicated in *how* law happens'.[77]

If landscapes, even ancient ones, become players in *how* law happens, in all its diverse performances,[78] then another potential source of evidence may be availed of. In the legal geography of Anglo-Saxon England, the forest and the fen (the coastal marshlands) were dominant physical features. Human settlements were scattered and small-scale: hamlets and farmsteads surrounded by village fields and expanses of waste, settled within expansive swathes of green carbon sinks.[79] The great forests of the Weald, the Essex–Chiltern belt, or the Bruneswald made 'about half of the southeastern quarter of England heavily wooded'.[80] In Warwickshire, the Forest of Arden, immortalised by William Shakespeare in *As You Like It*, dominated. And in the north, Sherwood Forest lay intact. Even in Norfolk, said to be 'one of the most highly developed counties' of the Anglo-Saxon age, there was sufficient woodland 'to mark off the centre from the rest of the county as a region in which swine were numerous and demesne sheep-farming was little developed'.[81]

Yet despite their vast footprint, few of these woods were untouched, pristine in their wilderness. Rather, as we have seen from the preceding discussion on common rights, they were rich community resources, providing:

> supplies of firewood, timber and game . . . and a source of food for sheep and pigs. Meadows were scanty in the extreme, the pastures consisted mostly of very rough grazing grounds, the insufficiency of which is reflected in the use of woodlands as swine-pastures.[82]

If co-constitutive of law, as legal geographers contend, these forested landscapes would have been implicated in a 'great body of familiar usages',[83] a

---

[76] Ibid.

[77] Ibid.

[78] Performance theory is also an aspect of legal geography, the idea that the law is enacted and does not exist in a prior state, Nicholas Blomley, 'Performing Property; Making the World' (2013) 27 *Canadian Journal of Law and Jurisprudence* 1.

[79] Windeyer describes 'England in 600 . . . [as] a land of marsh and thick forests, broken here and there by open spaces and clearings where Anglo-Saxon inhabitants lived in scattered agricultural communities', Windeyer (n 23) 1–2.

[80] Lennard (n 1) 250.

[81] Ibid 19.

[82] Ibid 3. Similarly resource-premised, the Anglo-Saxon wood was described by Peter Blair as 'much valued not only for its game, but also as swine pasture and a source of building materials and fuel', Blair (n 22) 251.

[83] Stephenson and Marcham (n 31) 1.

'making' of custom that reflected and protected the forest's natural resource values. Hence, in the Laws of Ine, the taking of trees by wanton felling or fire was wrongful, a grievous theft. But where their felling had a productive reason, the axe became an 'informer' not a 'thief'. Moreover, to harvest the forest's resources, villagers had to (logically) access the forest with few limitations. Again, this right may be assumed from the qualifications placed on 'those from afar', travellers who strayed off the highway into the woods. Theirs was a contingent entitlement, rights less extensive than those enjoyed by 'those who were near'. And finally, like the diversity of the woods themselves, such customs (and customary common rights) were commensurate to context – complex where the forests were vast, like the Forest of Arden and its arcane rules around timber use, but scant in places such as Norfolk, where (later) forest laws were never enacted since woods and glades were fewer.[84]

The Norman Legal Forest and the Forest Charter of 1217

While the Norman Invasion proved hugely disruptive to the Anglo-Saxon political order, it did not unduly disrupt the continuity of legal custom. Over a relatively short time, Norman laws adapted to the lie of the land. Certainly, the new regime marked a shifting of the legal axis – from an Anglo-Saxon to a hybridised Anglo-Norman landscape, and the beginnings of a distinctive common law – yet the change proved to be evolutionary, not revolutionary. Certain emergent trees may have been felled to the conqueror's axe, but the forest's root system and its 'underwood' lay largely undisturbed. One stark exception to this narrative of continuity was the Forest Law of the Norman Invaders, a (brief) legal disruption that *was* both revolutionary and highly instructive of what these laws sought to subvert and enclose – the heterogeneous diversity of forest custom.

William introduced the concept of Forest Law from continental Europe. Legal forests gave the king a right to hunt on lands he did not own, so-called 'afforested' lands where the rights of others, lay and ecclesiastical nobles but particularly commoners, were overridden or subjugated to the imperatives of the hunt. Forest law was deeply unpopular, a toxic political issue that intensified as their reach expanded under later Norman kings William II and Henry I, and especially (the first Plantagenet king) Henry II,[85] when it was estimated that afforested lands comprised one-third of the English landmass.[86] Under forest law, private rights to hunt enclosed customary rights, such that

---

[84] Suggesting, by implication, that any forest customs specific to Norfolk were likewise few.

[85] Hayman (n 68) 23.

[86] Judith Greene, 'Forest Laws in England and Normandy in the Twelfth Century' (2013) 86 *Historical Research* 416, 417.

common rights of pasturage, herbage or pannage, or rights to cut wood, were restricted or subject to 'fines'.[87] Despised foresters and their wardens meted out draconian punishments for transgressions against the 'vert' (the woods), but (especially) crimes against the 'venison' (the animals),[88] with poaching exacting cruel retribution through blinding, emasculation, or execution.[89] As Hayman says, 'Forest law came to be regarded as the acme of Norman despotism.'[90]

Forest law governed much more than forests. Indeed, the woods and their constituent trees were peripheral to what was an abstract legal artifice: a juridical space set aside for hunting and the exercise of royal privilege. The legal forest thus became detached from its treed context. Forests may have included woodlands, since woods were favoured haunts of deer (and other game), but as a legal device, the legal forest also encompassed forest clearings, villages, or open fields, 'great swathes of land held by others'.[91] Indeed, at their height, legal forests covered the entire county of Essex, or what became the 'New Forest', a creation of diverse land uses carved from the customary weald by William, and said via popular myth to entail 'the demolition of whole villages, the extermination of their inhabitants and the destruction of three dozen churches'.[92]

The Assize of the Forest of 1184 represented the high-water mark of the legal forest.[93] Through this instrument, Henry II forbade 'that any one shall commit any sort of offence, touching his venison and his forests'.[94] Henry outlawed 'bows, arrows, dogs or hounds',[95] guaranteed 'the rights of his knights or other good men' in their enjoyment of hunting privileges,[96] prioritised royal agistment,[97] and criminalised the 'waste' of his forests – while 'conceding that without waste others may take from the woods whatever may be necessary for them, *[but] this by view of the king's forester*'.[98]

---

[87] A form of (high) taxation.

[88] The term 'venison' encompassed a wide category of animals, including but not limited to deer.

[89] Hayman (n 68) 23.

[90] Ibid.

[91] Greene (n 86) 417.

[92] Ibid.

[93] It is also known as the Woodstock Assize.

[94] Art. 1, The Assize of the Forest (1184), in Stephenson and Marcham (n 31) 87–9.

[95] Art. 2. The latter was enforced through the 'lawing of mastiffs', the mutilation of a dog's feet so they could not chase game, Art. 14.

[96] Art. 6

[97] Art. 7. This especially restricted pannage, the putting of swine into the woods to feed, which four knights per county were authorised to police.

[98] Art. 3 (emphasis added).

Norman forest law was deeply unpopular from its inception. Historian Judith Greene argues that '[its] introduction represented one of the sharpest breaks in English society after 1066'.[99] For commoners, the list of forest 'inconveniences' included restrictions on pannage, 'the cutting or burning of wood [and] hunting, and the carrying of bows and spears in the forest'.[100] Access to the forest was also encumbered with new and onerous obligations: 'avoid[ing] letting animals loose in the forest, neglecting summonses, or not reporting those whom they met in forests with dogs, or discoveries of hides or flesh'.[101] In its ultimate articulation, the Assize of the Forest, with its arbitrary restrictions on what commoners could 'take' from the woods absent waste, was symbolic of executive hubris and overreach.

By the beginning of the thirteenth century, these 'inconveniences' were documented as the 'evil customs of the forest'. In 1215, the Magna Carta stipulated that 'twelve knights placed under oath in each county' investigate these evil customs, with the aim that action be taken 'within forty days' to 'abolish [them] completely and irrevocably'.[102] Two years later, its companion instrument, the Forest Charter, was enacted to deal specifically with abuses of Forest Law. Sealed on 6 November 1217, the Charter disafforested large tracts of land placed under the forest jurisdiction by Henry II.[103] Article 1 restored rights of access to the king's forests to all 'good and lawful men', along with rights of 'common of herbage and other things in the same forest to those who were accustomed to have them before'. Article 9 likewise restored rights of agistment and pannage, article 12 clarified that 'every freeman may with impunity make a mill, fish-preserve, pond, marl-pit, ditch or cultivate arable land' in the forest, while article 13 returned rights to 'have honey found in the woods'. However, the concluding article 17 was arguably the most sweeping, with its restoration of forest liberties ('the liberties and free customs in forests') granted 'to all persons of our kingdom, both clergy and laity'. Excluded from this grant, however, were the lay and ecclesiastically powerful – the 'archbishops, abbots, priors, earls, barons [and] knights'.

The Forest Charter is a remarkable statute for three compelling reasons: its universality, its longevity and its obscurity. The Charter's *universality* reflects what Finchett-Maddock calls its concern with 'socio-economic and common rights' – compared with the Magna Carta, which is the 'epitome

<hr/>

[99] Greene (n 86) 417.
[100] Ibid 420.
[101] Ibid.
[102] Art. 48.
[103] Arts 1 and 3.

of upholding individual rights'.[104] The Charter's 800th anniversary in 2017 provoked commentary that reinforced its (extraordinarily) egalitarian tendencies: it 'dealt with rights enjoyed by the common man rather than privileges of the barons',[105] 'it protected the lives and livelihoods of commoners from encroachment by the aristocracy', and 'it was genuinely for the common people ... [protecting] the historic rights of people to have access to the woods and royal forests, so they can carry on collecting firewood, feeding their livestock and foraging for food'.[106] Its granting of forest liberties *to all*, save the rich and powerful in article 17, underscores what a remarkable statute it was for its times, indeed *any* times.

The Charter's *longevity* reflects that it remained good law for over 750 years, replaced only in 1971 by the Wild Creatures and Forest Laws Act (UK), the briefest of Acts that nonetheless in sub-section 1(5) affirmed that 'no existing right of common or pannage originating in the forest law shall be affected by its abrogation'. And lastly, despite its remaining 'the statute in force for the longest time in England',[107] the Forest Charter is remarkable for its near-invisibility, an *obscurity* that Finchett-Maddock describes as being 'lost ... consigned to the chattels of history, very much like its subjects, the commoners and the common fields'.[108] This obscurity is unsurprising, as Chapter 1 canvassed, the Magna Carta's 'ridding of its communal twin' an example of 'the force and imposition of enclosure over the commons'.[109] Its near 'loss' reminds us of the pre-eminence of what Finchett-Maddock (similarly) calls a privileging of the memory of enclosure, and its concomitant relegation of the memory of the commons 'to antiquity through neglect'.[110]

---

[104] Finchett-Maddock (n 8) 129.
[105] Mark Hill, 'The Charter of the Forest 1217: A Medieval Solution for a Post-Modern Problem?' (Paper presented at the G20 Interfaith Summit, Istanbul, 16–18 November 2015), available at <https://www.iclrs.org/content/events/116/2377.pdf>.
[106] Fiennes (n 3) 123.
[107] Carolyn Harris, 'The Charter of the Forest', Magna Carta Canada (17 December 2013), available at <http://www.magnacartacanada.ca/the-charter-of-the-forest/>.
[108] Finchett-Maddock (n 8) 129.
[109] Ibid.
[110] Ibid 130. 'It can be said that the Charter of the Forest offers a similar depiction of a law of resistance, and this sententious link to the commons allows an image once again how they been [sic] wiped away, lost, but are still in existence somewhere outside of state law institutionalisation. It is in the memory-work of the commons that the Charter of the Forest is re-found, literally through the searching and researching of the archive of the commons, social centres, protest movements', ibid.

Conclusion

This chapter's allegory, its tale of the ancient forests of Anglo-Norman England and their ancient forest customs, ends abruptly in a nondescript statutory addendum to the Wild Creatures and Forest Laws Act of 1971, a provision that saves from abolition the 'rights of common or pannage originating from the forests'. This obscure sub-section typifies what we argue is the endurance and resilience of what lies at the bottom of Flach's woods. Subsection 1(5) may be the briefest of statutory edicts, yet it forms part of a vast, largely hidden jurisprudence of customary rights, an archived memory of the commons overwhelmingly relegated to antiquity – and obscurity – through its neglect. Our progression in one chapter from 1217 to 1971 (and onwards to today) is a hyper-speed journey through time, but not space. Like a visitation of Keenan's ghosts, we may not share a temporal dimension with the good and lawful men of the thirteenth-century forest, but we do inhabit the same (albeit dramatically altered) space, a legal geography of forest commons we can trace to the edge of legal memory. As Keenan eloquently explains, '[h]aunting is a way of thinking about how "now" is connected to "then"', of 'maintaining a relation of belonging with the place [the ghost] haunts'.[111]

This relation of belonging with place has been performed and replayed repeatedly over the centuries. Today, such ghostly apparitions materialise as infinitely diverse claims to place: resurgent celebrations of the winter solstice, Maypole dancing under the harsh Australian sun, foraging in public forests, or the protests of the Occupy movement. The power or legitimacy of these communal claims to place is drawn deep from the past – they are hauntings from the ghosts of the forest, from Flach's *fond des bois*. Sometimes these ghostly enactments 'hit you in the eye',[112] like the spontaneous infrastructure of Occupy with its tent cities, camp kitchens, yoga classes and libraries.[113] Mostly, however, these hauntings are subtle, nuanced – primarily registered in our subconscious minds, often brought to the fore through enactments of communal solidarity, or semi-meditative activities such as walking. The literature of walking, for example, hints at how certain pedestrians become absorbed in an 'eternity of consonance', an ethereal-like capacity to glimpse present sightings of the past.[114] In an analogous fashion, as seen in Chapter 1,

---

[111] Keenan (n 6) 169.

[112] Carol Rose, 'Seeing Property', in C M Rose (ed.) *Property and Persuasion: Essays on the History, Theory, and Rhetoric of Ownership* (Westview Press, 1994).

[113] Cristy Clark and John Page, 'Of Protest, the Commons and Customary Public Rights: An Ancient Tale of the Lawful Forest' (2019) 42 *UNSW Law Journal* 26.

[114] See, e.g., Frédéric Gros, *A Philosophy of Walking* (Verso, 2011)

the obscure field of psychogeography[115] seeks to articulate how geographical surroundings impact on a person's psyche, enabling an escape from the 'processes of banalisation' that reduce modern cities to a 'drab monotony' and instead to see 'the city as a site of mystery . . . to reveal the true nature that lies beneath the flux of the everyday'.[116] In London, Alfred Watson describes indecipherable 'ley lines' that connect across the metropolis, the word 'ley' derived, as we have seen, from the ancient idea of a forest clearing. There are 'zones within the city [London] which display chronological resonance with earlier events, activities and inhabitants'.[117] In Sydney, author and walker Vanessa Berry similarly writes of that city's 'hidden and enigmatic places':

> Cities are made up of histories and memories as much as they are made up of physical environments. As different as cities can be, all share the qualities of density and constant change. Traces of the eras they have passed through can be observed in their details and *felt* in the patchwork of atmospheres that shift from place to place.[118]

Such sightings of diverse claims to space – and the diverse performances of communal property that enact them – are enhanced through Berry's concept of the *radical potential of taking notice*.[119] Careful observation reminds us that ancient ways of *doing* remain near to hand. As David Dee observes, '[o]bjects, as it says on North American car mirrors, are closer than they appear. We fail to notice this when mediation is all.'[120] Dee's writings on place are apposite to property, which basically seeks to regulate our relationships with place (and space). Private property, with its domineering discourses of enclosure, commodity and alienation, mediates between people and place such that we perceive this private rhetoric to be all-consuming. Yet, as Dee also explains, places and their memories linger, 'stubbornly there, itchy, palpable, determining. Good or bad, they are felt under our skin, and get under it too.'[121]

---

[115] The genesis of the term 'psychogeography' is often attributed to French *situationistes* in the 1950s, a loose group whose subversive wanderings were designed to disrupt the hegemonies of urban space.

[116] Merlin Coverley, *Psychogeography* (Pocket Essentials, 2010) 13. Berry describes psychogeography as a 'heightened version of everyday engagement with the identities and moods of places. It gave a name to the act of examining the urban environment and connecting to the latent forces within it', Vanessa Berry, *Mirror Sydney: An Atlas of Reflections* (Giramondo Publishing, 2017) 8

[117] Cited in Coverley (n 116) 119–24.

[118] Berry (n 116) 4 (emphasis added).

[119] Ibid 210.

[120] David Dee, *Ground Work: Writings on Places and People* (Jonathan Cape, 2018) 2.

[121] Ibid 7.

These haunting memories speak to an *otherness* that is closer than it appears, a ghostly groundswell from the bottom of the woods – itchy, palpable and intermittently springing to performative life.

This chapter's objective is to describe the ancient forest liberties of long-gone ancient forests, and to remind us that its subliminal hauntings are not far from the surface of our modern, concreted-over places. It sees these ancient hauntings through imaginative inference, using a retrospective lens – a rear-view mirror, if you like – to imagine forest custom through the arcane language of the Forest Charter and the localised incidents of common rights. In seeing these visitations, this chapter raises provocations that go the very essence of *The Lawful Forest*. How has our 'memory of the commons' been relegated to the near-forgotten periphery? Where can we unearth the telltale signals of this memory of the commons? And, most significantly, what might be the *radical* implications of this reconceptualisation? Answers to these conundrums are critical, since it is the memory of the lawful forest that continues to haunt our modern-day relationships with land, an intuition that the forest stands as a symbol of our sense of belonging to place, grounded and utopian.

Chapter 3 next continues this (mostly linear) spatio-temporal journey. We stay in the constantly evolving space that was Anglo-Norman England, jumping another century or so 'forward' in time to an earlier pandemic, the devastating disruptions of the Black Death of the fourteenth century. This scarcely imaginable upheaval, unthinkable (we once thought) to modern minds, set the stage for the three centuries that followed, the possibilities of alternate fork(s) in the forested path, and the glimpsed utopias that shimmered briefly yet brightly in their unrealised promise.

3

A Glimpsed Utopia

GLASTONBURY THORN TREE

In this chapter, we jump forward to the fourteenth to sixteenth centuries, to provide a historical account of the early enclosures of rural (and urban) commons, and resistance to this enclosure via a Peasants' Revolt, sporadic riots and a fictional Utopia. In this chapter, the notion of utopia is explored and defined through historic example and theory. It is used to further explore the prefigurative politics of forests, and some of the history behind the experimental communities and protest camps considered in later chapters.

This chapter begins with a brief exploration of the manorial system that emerged around the time of the Norman Conquest and the changes that were wrought to this system by the population shifts caused by the Black Death. From here, the chapter explores some of the key rebellions that occurred in the fourteenth and fifteenth centuries, including the Peasants' Revolt of 1381, the early urban enclosure riots and the radical theology that informed these uprisings. This exploration of radical theology then leads to a consideration of Thomas More's *Utopia*, a literary work of both fiction and socio-political satire that critiqued the times and portrayed a place (or rather 'no place') where enclosure was unthinkable and communal ownership of property the norm. More's utopian thought experiment then gives way, in Chapter 4, to Kett's Rebellion, the English Civil Wars, the Levellers and the True Levellers (or Diggers) – and Gerrard Winstanley's concrete utopian experiment. Through these various incidents of both theory and action, this chapter tracks the history to what has been called England's second or 'shadow' revolution[1] – one that focused on the rejection of enclosure, claims to communal property rights, and radical economic equality in lieu of Cromwell's bourgeois republic that was eventually realised.

---

[1] Christopher Hill, *The World Turned Upside Down: Radical Ideas during the English Revolution* (Penguin Books, 1975).

The Manorial System and the Black Death

As Chapter 2 depicts, early 'Anglo-Saxon settlements were relatively small, and surrounded by wood and waste'.[2] These settlements of 'individual free peasant landholders' came to symbolise a particular conception of pre-Norman English life that was subsequently held up in contrast to the strictures of the so-called 'Norman Yoke' – a symbolic concept that came to play a powerful role in the rebellions of later centuries. Historians, however, dispute this idealised version of English history and point out that 'the general drift of peasant life in [the] centuries before the Norman Conquest was from freedom to servitude'.[3] According to Rowland Prothero, one reason for this shift was that the law at this stage of English history 'was powerless to protect individual independence or to safeguard individual rights'.[4]

As a result of these shifts, the 'self-governing, independent communities of freemen [who] originally owned the land in common . . . were gradually reduced to dependence by one of their members, or by a conqueror, who became lord of the soil'.[5] This arrangement meant that most medieval agriculture, like other industries, came to be a collective practice organised around the 'manor'.[6] Frank Stenton argues that while records are insufficient to determine exactly when the manorial system was instituted, it is clear that the evolution of a manorial economy began 'many generations before the Conquest'.[7]

Change came to England and the manorial system in a sudden and gruesome manner, in the form of the Black Death. The English population was reduced by half, from around 5 million in 1348 to 2.5 million in 1377, and did not recover until after 1520.[8] In this era, the population and its material wealth contracted, and 'most villages and towns bore the scars of collapsed buildings, and settlements as a whole or in part were abandoned'.[9] Although these scars on the landscape represented a kind of dystopia, a

---

[2] George Edwin Fussell, 'Introduction', in R E Prothero, *English Farming: Past and Present* (Cambridge University Press, 6th ed., 1961) xxxviii.

[3] Frank Stenton, *Anglo-Saxon England: Volume 2 of Oxford History of England* (Clarendon Press, 3rd ed., 1971) 470.

[4] Rowland E Prothero, *English Farming: Past and Present* (Cambridge University Press, 1912) 3.

[5] Ibid 4.

[6] Ibid 3.

[7] Stenton (n 3) 480.

[8] Christopher Dyer, *Making a Living in the Middle Ages: The People of Britain 850–1520* (Yale University Press, 2003) 265.

[9] Ibid.

significant result of this declining population was a shortage of labour and tenants, which shifted the balance of power away from the landed gentry.[10] The price of agricultural produce was also low during this period, and there was a shortage of coin (especially of small denominations).[11] As a result, Christopher Dyer describes this period as 'a time of liberation, when old restraints were removed and new freedom of choice opened'.[12] Here we can see one of history's forks in the road, where new possibilities for social and economic arrangements opened up due to a break in the norms and a shift of power between the elite and the masses.

During this period, serfs claimed freedom from their lords' manors – often by simply leaving without permission and successfully seeking work elsewhere.[13] Labourers also resisted historically exploitative arrangements by refusing to accept long-term contracts and negotiating higher wages than those that were established under the Statute of Labourers of 1351.[14] Acute labour shortages opened up many new employment options to free labourers, including the opportunity to purchase licences to live within the walls of cities and to find town employment.[15] The Labour Statutes were even relaxed enough to permit immigration to London.[16] Finally, the reduced population enabled many smallholders to acquire extra land or move to larger holdings – a practice called *engrossment*.[17]

In response to this emerging lower-class power, the upper classes relied on the legal system to reassert their historically entrenched privileges and to reimpose existing social controls. This response began with the Parliamentary Commons, which shifted from its historic tendency to give voice to the grievances of the peasantry against the royal government in matters such as purveyance (the right of the Crown to requisition goods and services for royal use), the collection of wool levies, and the high cost of judicial writs.[18] In 1351, the Commons successfully petitioned for the Statute of Labourers, 'which attempted to peg wages at pre-plague rates'.[19] As John Maddicott recounts:

---

[10] Ibid 266.
[11] Ibid.
[12] Ibid 278.
[13] Ibid.
[14] Ibid 278–9.
[15] Prothero (n 4) 52.
[16] Ibid 53.
[17] Dyer (n 8) 279.
[18] J R Maddicott, 'Parliament and the People in Medieval England' (2016) 35(3) *Parliamentary History* 336, 343–5.
[19] Ibid 354, citing *The Parliament Rolls of Medieval England, 1275–1504* (PROME) v14, 28–30.

Then followed a series of further repressive petitions put forward by the Commons through the 1360s and 1370s. In 1368 they asked that labourers and artisans demanding what they called 'excessive and outrageous prices and salaries' should be made to pay back twice the excess above the statutory level. In 1372 they complained about employers who received labourers fleeing from their own districts to great towns or other regions and who paid excessive wages to the fugitives. In 1376 they asked that fugitive labourers leaving their manors to look for work elsewhere should be put in the stocks or imprisoned. And in 1377, during Richard II's first parliament, they requested that labourers should be forced to work for designated employers who had a claim to their labours.[20]

In a comparison with the responses of other European states to the disruptions of the Black Death, Samuel Cohn notes 'England's steadfastness in its imposition against rural labour'.[21] As examples, he lists a series of labour laws that introduced 'stiffer penalties on rural labourers and curtailed their social and geographic mobility with harsher controls'.[22] Edmund Fryde similarly documents a thirty-year period after the Statute of Labourers during which the Commons petitioned for a raft of repressive measures against the peasantry, although he does note that many of these petitions were rejected by the royal government.[23] He also argues that although the Statute was 'perhaps the most zealously enforced ordinance in Medieval English history',[24] with court records documenting heavy reliance on it to charge workers for taking high wages or for 'wandering from village to village "for excess"',[25] it was actually 'fairly ineffective'.[26] Indeed, 'employers found it to their advantage to ignore it and could afford to do so',[27] indicating that it may have been more cost-effective to simply pay the fines than to comply. This may partly explain why labour laws in 1361 replaced monetary fines with imprisonment and branding.[28]

---

[20] Maddicott (n 18) 354.

[21] Samuel Cohn, 'After the Black Death: Labour Legislation and Attitudes towards Labour in Late-Medieval Western Europe' (2007) 60(3) *Economic History Review* 457, 476.

[22] Ibid.

[23] Edmund Fryde, 'Peasant Rebellion and Peasant Discontents', in E Miller (ed.) *The Agrarian History of England and Wales, III: 1348–1500* (Cambridge University Press, 1991) 744, 759.

[24] Ibid 755.

[25] Dyer (n 8) 283.

[26] Fryde (n 23) 758.

[27] Ibid.

[28] Cohn (n 21) 476.

Dyer argues that during this time, 'the ruling groups temporarily closed ranks, and used the power of the state to defend the interests of the rich in a blatant manner'.[29] Just as the Occupy protests were spurred on by a general sense that state institutions had been captured by the 1 per cent,[30] Dyer argues that '[t]he ranks of society below the gentry felt that the state was losing any claim to impartiality as it became so closely identified with the landed interest.'[31] In addition to using the law to suppress wages and labour power, the landed gentry also used the manorial courts to increase their revenues via the imposition of petty fines.[32] A stark lesson from this period was the strength of the link between property and power, with spatial inequality fuelling broader aspects of injustice, and power disparity in turn reinforcing and entrenching spatial inequality. Those without property were denied liberty, particularly freedom of movement and the right to bargain for fair wages, which may help to explain why later philosophers came to conflate these two values. But there was nothing natural in this link between property and liberty. It was always a political choice – and a contested one at that.

The Lollards and the Peasants' Revolt of 1381

Religion was a key arena for this contest of political legitimacy in the late fourteenth century, particularly due to the growing popularity of radical theology. An influential figure in this movement was John Wycliffe, a University of Oxford philosopher and theologian (known also as Wyclif and as the 'Morning Star of the Reformation'). In addition to rejecting the doctrine of transubstantiation, Wycliffe promoted the idea that common people's faith should be guided by their own reading of the Scriptures.[33]

Wycliffe is perhaps best known for translating the Bible into the vernacular of the day, believing that 'the faith of the church is contained in the Scriptures, [so] the more these are known in a true sense the better'.[34] This radical emphasis on the importance of a personal relationship with the Scriptures undermined the power of the Pope and the clergy, which made Wycliffe particularly unpopular with the Pope in Rome. In 1378, he was

---

[29] Dyer (n 8) 285.

[30] David Graeber, *The Democracy Project: A History, a Crisis, a Movement* (Random House, 2013).

[31] Dyer (n 8) 285.

[32] Ibid.

[33] See, e.g., Margaret Aston, 'John Wycliffe's Reformation Reputation' (1965) 30 *Past & Present* 23, 25–6, 36.

[34] Christopher K Lensch, 'The Morningstar of the Reformation: John Wycliffe' (1996) 3(2) *WRS Journal* 16.

charged with heresy, and not long after forced out of Oxford. Nonetheless, he continued to write and preach until his death in 1384.

Wycliffe's followers, Wycliffites, were labelled 'Lollards' – a pejorative term meaning 'mumbler' or 'mutterer' – and their theology, along with similar beliefs that were circulated at the time, helped to foment dissent within the lower classes throughout this period. As William Pelz has argued:

> Any call to resist papal exploitation, such as church taxes, quickly led people to question secular oppression. If the Church had strayed from the teachings of Christ, should not Christians live as the Son of God in the same manner his disciples had? If, as Jesus said, 'It is easier for a camel to go through the eye of a needle than for a rich man to enter the kingdom of God,' how should we think of the nobility? And, if Christ is the model, why not share everything as had been the practice of Jesus and the early believers. When commoners learned from readings from Acts of the Apostles the manner of Christian life in past times, some found inspiration to reject the feudal order. Imagine a European peasant hearing for the first time that Christians had held 'all things in common,' and any surplus was 'distributed to each as anyone had need.'[35]

These simmering class tensions came to a head in 1381 with an unprecedented peasants' rebellion, or rebellion of the *comuynes* if we account for the 'occupational heterogeneity of the movement'.[36] In counties across south-east England, the 'labouring classes' – or '*laboriaris*'[37] – revolted: 'They attacked lawyers, abbots, tax-collectors, and royal commissioners; they burned title-deeds and manor rolls, broke open jails and liberated prisoners, occupied Canterbury, St. Albans, St. Edmundsbury, and Norwich, and marched on London.'[38] It is notable that manor rolls were targeted in this record-burning spree, as they were the indicia of both private property and the related seigneurial restrictions on tenants or *villeins* – the central cause of their frustrated aspirations for a better life.[39] As Herbert Eiden argues, this implication 'is clearly revealed by the first court after the burning of the records in

---

[35] William A Pelz, *A People's History of Modern Europe* (Pluto Press, 2016) 21 (references omitted).

[36] David Rollison, 'The Specter of the Commonalty: Class Struggle and the Commonweal in England before the Atlantic World' (2006) 63(2) *The William and Mary Quarterly – Class and Early America* 221, 234 ('Specter of the Commonalty').

[37] Ibid 222.

[38] Anna Vaninskaya, 'Dreams of John Ball: Reading the Peasants' Revolt in the Nineteenth Century' (2009) 31(1) *Nineteenth-Century Contexts* 45, 45.

[39] Herbert Eiden, 'Joint Action against "Bad" Lordship: The Peasants' Revolt in Essex and Norfolk' (1998) 83(269) *History* 5, 22–3.

Wivenhoe (Essex) in January 1382', with the records indicating that the tenants responsible 'claimed "to hold the said tenements at their own will, freely, and not at the will of the lord"'.[40]

Chronicles of the events differ in their details, but it is broadly agreed that the final straw was the imposition of a third flat-rate poll tax within a four-year period and widespread discontent with the perceived incompetence and fraud within government.[41] The preceding decade had been marked by poor governance, first from a senile king (Edward III) and then from those who took charge during his grandson's minority (Richard II).[42] Particular animosity was felt towards those individuals in central government who were held responsible for mismanagement and corruption, including John of Gaunt ('who effectively controlled the realm at this time'[43]), Simon Sudbury (royal chancellor and Archbishop of Canterbury) and Robert Hales (royal treasurer and Prior of the Knights Hospitallers).[44] All this came on top of the simmering class tensions that stemmed from the demographic and economic changes brought about by the Black Death, the subsequent labour laws, and the drawn-out effects of the Hundred Years War (particularly including its drain on revenue and the related abuses of purveyance).[45]

Indeed, Eiden argues that examining the social and economic standing of the insurgents (at least those from Essex and Norfolk) provides a valuable insight into both the motivations and the identities of the rebels.[46] In both counties, 'a striking feature . . . is the prominence of the insurgents in the administration of their respective manors or hundreds either before the rising or shortly afterwards', with a significant number serving as 'assessors or sub-collectors of one of the three poll taxes, as constables, bailees, reeves', and so on.[47] While peasants were certainly represented, these rebels came from a wide range of occupations, 'from the aspiring ranks of labourers and craftsmen as well as from the middle and upper levels of the tenantry',[48] with the textile and clothing sector being strongly represented.[49] Eiden notes that

---

[40] Ibid 29, citing Essex Record Office, T/B 122.
[41] Dyer (n 8) 287; Andrew John Prescott, 'Judicial Records of the Rising of 1381' (PhD thesis, Bedford College, University of London, 1984) 14.
[42] Prescott (n 41) 14.
[43] Ibid, citing Anonimalle Chronicle, 134–5; Knighton, ii, 130–1; Eulogium Historiarum sive Temporis, ed. F Haydon, iii (Rolls Series, 1863), 351–2, but cf. 127–9.
[44] Dyer (n 8) 287; Prescott (n 41) 14.
[45] Eiden (n 39) 7.
[46] Ibid 26.
[47] Ibid.
[48] Ibid 28.
[49] Ibid 26.

'[a] significant minority [of the rebels in Essex and Norfolk] held land on disadvantageous terms' and reasons that '[t]he participation of the village elite exercised a decisive influence upon the coherence and organization of the revolt.'[50] He goes on to argue:

> The implications for the interpretation of the revolt are considerable –
> especially since it can be shown that some of the labourers and craftsmen
> are referred to in the manorial court rolls as native or customary tenants
> with obligations to serve on their lord's demesne. A good deal of their dis-
> satisfaction may have derived from the lack of fit between their commercial
> activities on the one hand and their seigneurial subordination on the other.
> Many lords at this time still adhered to the way they or their ancestors had
> treated the peasantry in pre-plague days, despite the social and economic
> changes of the 1370s and 1380s. The dynamics of this process of change
> had made the potentially independent craftsmen and peasants more self-
> confident. The old manorial regime had had its day, but the lords of the
> manor had not realized it yet.[51]

Also implicated in the rebellion was one John Ball, radical preacher and contemporary of Wycliffe, who had long called for economic redistribution. As Pelz points out, '[s]ince Christianity pervaded European language and culture at that time, the rebellion, too, would be phrased in Biblical tones even if, for some, the motivation may have been a bit more secular.'[52] These links between the democratisation of religion and the radicalisation of the masses underscore the key functions of the state Church in controlling the population, legitimising the current order and quelling dissent. Justifying the existing system, with its entrenched spatial inequities, required a selective reading of the Scriptures. This was most easily achieved when the Bible was inaccessible to the masses – although, as modern-day fundamentalists have demonstrated, this is not necessarily required.

While the rebellion resulted in uprisings across England, some of the rebel bands headed for London, led by the commoner Wat Tyler.[53] When

---

[50] Ibid 28–9.
[51] Ibid 28.
[52] Pelz (n 35) 21 (references omitted).
[53] Dyer (n 8) 287. Little is known about Wat Tyler's history or life, but a literary account of his role in the Revolt developed a life of its own over subsequent centuries – serving first as a cautionary tale for those who would presume to challenge the king's divine prerogative, and then as a hero and source of inspiration for later revolutionary movements, including the English Revolution, the Chartists and the American Independence Movement – see Stephen Basdeo, *The Life and Legend of a Rebel Leader: Wat Tyler* (Pen and Sword, 2018).

they reached London, they destroyed property including Gaunt's palace of the Savoy, the Temple Bar and the headquarters of the Knights Hospitallers, and captured and executed a number of persons, including Hales and Sudbury.[54] The following day, the rebels demanded to see the young king, Richard II, and met him to the east of the city at Mile End.

The details of the demands presented to the king are the subject of some debate. Some chronicles document the rebel demands as including 'a detailed and wide-ranging programme' to abolish serfdom and forced labour and reduce taxation, and the granting of a broad amnesty to all rebels.[55] Others claim that the rebels made 'less elaborate demands' focused on 'personal freedom and a pardon for any offences committed in the course of the insurrection'.[56] Most historians seem to agree that they had a more radical structural agenda.[57] Andrew Prescott, for example, concludes that '[w]hatever the precise nature of the demands made at Mile End, it is clear that the insurgents pressed strongly for the abolition of serfdom and customary tenure.'[58] Dyer reports that the list of demands 'appears to have envisaged the removal of aristocratic privilege and the church hierarchy, and the creation of a popular monarchy in which the king ruled over self-governing village communities'.[59] The king appears to have granted at least some of the rebels' demands by issuing charters of emancipation and amnesty and agreeing to the removal of certain 'traitors'.[60]

The Revolt ultimately ended without immediate success. Tyler was stabbed by the mayor of London, William Walworth, and the rebels dispersed.[61] While a number of local rebellions continued across England, they were ultimately suppressed, with key participants being tried and executed.[62] Parliament also annulled the charters of emancipation and pardon.[63] Nonetheless, Maddicott argues that the Revolt 'delivered a powerful and salutary jolt to the basic assumptions of the parliamentary classes about the essential passivity of the lower orders'.[64] He quotes the Parliamentary Commons as blaming the 1381 revolt on

[54] Vaninskaya (n 38) 45.
[55] Prescott (n 41) 12–13.
[56] Ibid.
[57] Vaninskaya (n 38) 45.
[58] Prescott (n 41) 3.
[59] Dyer (n 8) 287–8.
[60] Vaninskaya (n 38) 45; Dyer (n 8) 287–8.
[61] Vaninskaya (n 38) 45.
[62] Vaninskaya (n 38) 45; Dyer (n 8) 287–8.
[63] Vaninskaya (n 38) 45.
[64] Maddicott (n 18) 346.

'the grievous and outrageous oppressions inflicted on [the poor commons] by the ministers of the king and of the other lords of the realm', and of the 'outrages ... newly committed against the poor commons more generally than ever before', for which reason 'the lesser commons rose up and wrought havoc in the said tumult'.[65]

According to Maddicott, the Commons ultimately vindicated the apparent assumption of the rebels that direct action was the most effective means of changing the system and increasing justice, 'not only by their attempts to suppress corruption and lawlessness in the wake of the revolt but also through their refusal to countenance any further grants of taxation in the years immediately after 1381'.[66]

The rebels had sought to take advantage of the rupture created by the plague to call for society to take a different path. They fought for an imagined alternative future in which spatial justice and popular sovereignty would take primacy, and both property and power would be more equally distributed. In response, Parliament made relatively minor concessions by clamping down on only the worst excesses of the upper classes. Nonetheless, David Rollison believes that a new kind of identity and constitutional norm was forged in the rebellion of the *comuynes* – a new foundational expectation that those in power would serve the 'commonweal' and be answerable to the people should they fail to do so.[67] As he concludes, '[t]he comuynes of 1381 were the first to raise the banner of popular sovereignty. It would prove to be a tenacious cause.'[68]

Early Enclosure

The same population changes that fomented the rebellion also contributed to the first wave of enclosure, which began around the late fourteenth century, with wealthy landlords appropriating and fencing off land that had previously been held in common by villagers.[69] Strictly speaking, enclosure was a breach of the common law, but it was difficult to enforce the law against the landed gentry as they tended to control the courts.[70] Rowland Prothero argues that although changes from this first wave of enclosures came at the

---

[65] Ibid.

[66] Maddicott (n 18) 347.

[67] Rollison, 'Specter of the Commonalty' (n 36); David Rollison, *A Commonwealth of the People: Popular Politics and England's Long Social Revolution, 1066–1649* (Cambridge University Press, 2010) (*Commonwealth of the People*).

[68] Rollison, 'Specter of the Commonalty' (n 36) 235.

[69] Lawrence Wilde, *Thomas More's Utopia: Arguing for Social Justice* (Routledge, 2017) 38.

[70] Ibid; Nicholas Blomley, 'Making Private Property: Enclosure, Common Right and the Work of Hedges' (2007) 18(1) *Rural History* 1, 4 ('Making Private Property').

cost of prosperity, the loss of land for the newly landless labourers 'was the means and price of . . . personal emancipation'.[71] He argues their smallhold-ings were largely surrendered voluntarily as a result of below-subsistence yields, rather than being forcibly claimed for sheep farming.[72] While many also opted to continue to work as agricultural labourers, their earnings were threatened by the conversion of tillage to pasture.[73]

Although these changes did increase freedom of movement, they also reflected a new level of precariousness for farm labourers and an increasing move from subsistence to commercially driven agriculture.[74] We can observe present-day parallels here with the move to 'flexible working arrangements' and the 'gig economy' being sold as a pathway to increased freedom for workers despite precarity and freedom of capital being the real winners.[75] As the pace of the enclosure movement increased, further social consequences became evident. By the late fifteenth century, the Tudor enclosure move-ment, with its promotion of large-scale sheep farming, 'was a major con-tributory factor to the desertion of two thousand villages between 1370 and 1520'.[76] The depopulation of rural areas was widely documented in 'litera-ture, pamphlets, doggerel ballads, sermons, liturgies, petitions, preambles to statutes, Commissions of Enquiry, [and] Acts of Parliament'.[77] The result was the widespread 'decay of farm-houses and cottages, loss of employment, eviction of tenants, [and] rural depopulation'.[78] Again, these dramatic visible consequences of economic shifts have modern parallels, with the hollowing out of cities like Detroit being an obvious example.

As a result of these wider ramifications, this early enclosure movement was soon met with backlash. In 1484, the lord chancellor called the attention of Parliament to the rural exodus caused by enclosure.[79] From 1489 onwards, this call was followed by many Acts of Parliament that sought to both halt and even undo the process of enclosure.[80] In the early sixteenth century, the lord

---

[71] Prothero (n 4) 52.

[72] Ibid 52–3.

[73] Ibid 53.

[74] Ibid 57–8.

[75] See, e.g., Robert MacDonald and Andreas Giazitzoglu, 'Youth, Enterprise and Precarity: Or, What Is, and What Is Wrong with, the "Gig Economy"?' (2019) 55(4) *Journal of Sociology* 724; Rebecca E Zietlow, 'The New Peonage: Liberty and Precarity for Workers in the Gig Economy' (2020) 55(5) *Wake Forest Law Review* 1087.

[76] Wilde (n 69) 38.

[77] Prothero (n 4) 58.

[78] Ibid 59.

[79] Ibid 60.

[80] Ibid.

chancellor, Cardinal Thomas Wolsey, 'personally invested himself in enforc-
ing obedience to the laws against the decay of houses and farm-buildings and
against the conversion of arable lands to pasture'.[81] Ultimately, however, none
of these official responses were successful in stopping its inexorable march.[82]

Early Enclosure Riots

A less official response came from the people themselves, who participated in
a wave of anti-enclosure riots and other forms of opposition from the late fifteenth century.[83] While many of these took place in rural areas, there was an
early wave of urban enclosure riots between 1480 and 1525 in towns across
England, including York, Coventry, Nottingham, Colchester, Southampton,
Gloucester and London.[84] Christian Liddy explains that, in contrast to the
later rural enclosure riots, these urban riots were less about land or material
goods than they were 'about the infringement and loss of collective rights'.[85]
He argues that 'the concept of citizenship was viewed through a spatial lens,
which made urban dwellers highly sensitive to any encroachment upon communal space'.[86] Emphasising the geographic nature of this political contest,
this particular 'notion of citizenship was grounded, quite literally, in the
soil'.[87]

In this context, conflict over the urban commons reflected the symbolic
meaning of the commons themselves – 'they were the terrain upon which
wider political struggles were fought'.[88] This was further emphasised by the
fact that the word 'commons' was used to mean both the common *people* (i.e.
commoners or the constitutional class of 'those who work'[89]) and their communal *property*. These wider struggles focused particularly on questions of
power and 'the degree of popular inclusion in town government'.[90] In a scene

[81] Ibid 60–1.
[82] Wilde (n 69) 38, citing John Thomson, *The Transformation of Medieval England, 1370–
      1529* (Longman, 1992) 41–6.
[83] Christian D Liddy, 'Urban Enclosure Riots: Risings of the Commons in English Towns,
      1480–1525' 226(1) *Past & Present* (2015) 41; Briony McDonagh, 'Making and Breaking
      Property: Negotiating Enclosure and Common Rights in Sixteenth-Century England'
      (2013) 76 *History Workshop Journal* 32; Roger B Manning, 'Patterns of Violence in Early
      Tudor Enclosure Riots' (1974) 6(2) *Albion* 120 ('Patterns of Violence'); Blomley, 'Making
      Private Property' (n 70) 4.
[84] Liddy (n 83) 41–2.
[85] Ibid 44.
[86] Ibid 45–6.
[87] Ibid 46.
[88] Ibid.
[89] Rollison, 'Specter of the Commonalty' (n 36) 221–2.
[90] Liddy (n 83) 46.

that pre-echoed Trump's tweet-rage against the Mayor of Washington DC over Black Lives Matter protests in the city,[91] the deeper political significance of these urban enclosure riots was hinted at in the reported reaction of Henry VII, who summoned the mayor of York to Greenwich in February 1495 to warn him to get his city in order, or 'I most and woll put in other rewlers that woll rewle and govern the Citie accordyng to my lawez'.[92]

Liddy argues that '[t]he maintenance of town commons was [considered to be] a barometer of good government.'[93] Reflecting this understanding, commoners employed both the language of rights and the concept of the 'common weal' to imbue their claims (and even their riots) with a powerful sense of legitimacy.[94] In Southampton, for example, the commons chided their rulers for allowing their common land to be enclosed, arguing that it was '"ageynst the Comyn weele"', while commoners in Saltmarsh justified their tearing down of the hedges enclosing land around the city by arguing 'that the Saltmarsh had been "occupyed for a Commyn wele" for all the freemen'.[95]

Rollison argues that from the time of the 1381 Revolt until 1649, this concept of the *commonweal* 'defined the constitutional culture of the age'.[96] Here,

> the *wele* of the commune was to be measured by the condition of its poorest, least prestigious members. If all the institutionalized layers of the ruling classes failed in their duty to serve the *wele* of the community, it was the duty of the *comuynes* to rise in its defense.[97]

Thus, it was a term that raised the spectre of communism, or 'of resistance and rebellion that welled up below . . . and got out of control. By how fuzzy a chain of associations, the reasons for rebellion could merge one great reason: the abolition of all institutionalized inequalities.'[98]

This link between the commons, political power and questions of communal *rights* is one that comes up again and again in the history of struggles

---

[91] @realDonaldTrump (Twitter, 30 August 2020, 7:08am) <https://mobile.twitter.com/real DonaldTrump/status/1300027733293047810>, '@MayorBowser should arrest these agitators and thugs! Clean up D.C. or the Federal Government will do it for you. Enough!!!'

[92] York Civic Records, ed. Angelo Raine, 8 vols (Yorkshire Archaeological Society, record ser., xcviii, ciii, cvi, cviii, cx, cxii, cxv, cxix, Wakefield and York, 1939–53) (hereafter YCR) ii, 115–16, cited in Liddy (n 83) 41.

[93] Liddy (n 83) 58.

[94] Ibid. See also Rollison, *Commonwealth of the People* (n 67).

[95] Liddy (n 83) 58. See also Rollison, *Commonwealth of the People* (n 67).

[96] Rollison, 'Specter of the Commonalty' (n 36) 245.

[97] Ibid 239.

[98] Ibid 245.

for spatial justice. Communal property is not only a source of livelihood and economic power, it is also an important site for participating in public life and strengthening the relationships within the commons (as a populace) and between them and their commons (the land). Furthermore, the rituals associated with these commons are an important link to the past and to the narratives that shape identities. As Liddy points out in his writing about the urban enclosure riots, '[c]ommon lands gave citizens not only a sense of place but a means of remembering the past.'[99]

One such ritual was the symbolic practice of walking or 'riding the bounds', or 'riding the franchise' as it was also known.[100] This was an early example of Nicholas Blomley's 'performing property and making the world',[101] but here (remarkably, at least to modern eyes) the spatial claim was communal, not individual. In later centuries, private property mimicked this practice with land owners walking the 'metes and bounds' of their lands. Yet, remarkably again, communitarian performances still linger, like public rights of way that access beaches in the United States, created by prescription – the open, peaceful, non-consensual action of walking the same path over twenty or more years. In some towns, the riding of the bounds took place on the same day that a new mayor swore the oath of office, and while this was certainly symbolic of the new mayor's position and legitimacy, riding the city limits also allowed the community to assert their rights as citizens and their claims over the town commons.[102] As Liddy argues:

> Town liberties were conceived spatially. This physical conception was, in turn, shaped by the common lands to which the citizens demanded access. The boundaries of town commons and urban franchises were frequently coterminous. Citizens thus viewed common lands as an integral aspect of the town's corporate liberties that they swore to uphold at their enfranchisement.[103]

Here we can see that two intersecting narratives, or contests, were being pursued. The first focused on the role of the common people in the government of the day and their broader citizenship rights in the life of the city (or the village) in which they resided – their '"[f]ranchise", "liberty" and

---

[99] Liddy (n 83) 56.
[100] Ibid 54. The 'franchise' is equally an obsolete common law property right.
[101] Nicholas Blomley, 'Performing Property, Making the World' (2013) 27(1) *Canadian Journal of Law and Jurisprudence* 23.
[102] Liddy (n 83) 54.
[103] Ibid 55 (citations omitted).

"freedom"'.[104] The second contest was specifically around claims to property – and, more significantly, around the meaning and purpose of property. During this era, property was still understood 'as a bundle of overlapping and often non-exclusive rights and obligations',[105] and the task for the enclosers was to establish a new understanding of property as a private and territorially bounded concept and to assert the legitimacy of this conception.[106] Blomley describes this emerging conception of property as '[i]ncreasingly disembedded from local social conditions, . . . [making it] possible to treat property as an abstract relation between an owner and a thing rather than a conditional tenure, associated with obligations to others.'[107] This was a gradual process, however, and 'the formal legal recognition of absolute property was slow in coming'.[108] Indeed, as mentioned earlier, Manning argues that 'no clear and unqualified definition of "property" can be found in any legal dictionary or the works of any legal writer before the eighteenth century'.[109] Struggles over enclosure were thus struggles over the legitimacy of this new attempt to construct property as private and commodified, and over the implications of this construction for the commons (in both senses of the word).[110]

In the rural context, commoners used a wide variety of strategies to resist enclosure – including litigation and direct action – to assert their customary rights to the commons.[111] However, many of these same strategies were also used by landlords to both maintain and assert the legitimacy of their newly established private claims.[112] During this period, many of the cases were litigated in the Star Chamber – or the chancery court of equity – which 'was clearly recognized by contemporaries as a forum in which title to property could be decided'.[113] In reviewing 800 of such cases, Briony McDonagh reports:

> in pursuing their cases through the courts [enclosers] tried to assert private property rights over land which had arguably once been subject to common

---

[104] Ibid 54.
[105] Blomley, 'Making Private Property' (n 70) 5.
[106] Ibid; McDonagh (n 83) 32–56.
[107] Blomley, 'Making Private Property' (n 70) 1–21, 2–3.
[108] Ibid 4.
[109] Roger B Manning, *Village Revolts: Social Protest and Popular Disturbances in England, 1509–1640* (Oxford University Press, 1988) 5, cited in Blomley, 'Making Private Property' (n 70) 4.
[110] Liddy (n 83); McDonagh (n 83) 32–56; Manning, 'Patterns of Violence' (n 83); Blomley, 'Making Private Property' (n 70) 4.
[111] McDonagh (n 83).
[112] Ibid.
[113] Ibid 35.

rights. At the same time, commoners might utilize the courts to protect their customary rights and reverse enclosures.[114]

The record also indicates a complex relationship between litigation and direct action, and that,

> rather than simply being a pretext for litigation or an escalation to violence, the direct action reported in the Star Chamber suits was part of a process of negotiation – over property rights, land use and the meaning of property itself – taking place both within the court and outside it.[115]

Direct action itself was varied and included tactics such as hedge breaking, impounding and rescuing animals, continuing to collect historically common resources, and mass ploughing. Of these tactics, hedge breaking (or 'levelling' as it also came to be known) is perhaps the best known and most symbolic of the era. Blomley argues that hedges were a crucially important 'device through which new forms of spatial discipline were both materialised and enforced'.[116] As Blomley notes, hedges were at the centre of 'a fierce political struggle over enclosure and privatisation' – they 'both helped to concretise a new set of controversial discourses around land and property rights, and aimed to prevent the forms of physical movement associated with the commoning economy'.[117] As a result, the conflict around hedges was both practical and symbolic. Just as hedges acted as both a physical barrier to entry and a signal of the enclosers' claims to exclusive possession, hedge breaking was a concrete means of enabling people and animals to enter property to continue with their customary use of the land and a symbolic reassertion of common rights.[118] As McDonagh argues, '[t]hat hedges were not only dug up but also burnt and buried draws attention to the considerable time and effort which was invested in hedge-breaking, as well as to the symbolic or ritualistic aspects of enclosure opposition.'[119] Like so many performances of 'proprietorial resistance'[120] prior to and hence, hedge breaking was at once physical and metaphysical, abstract and material.

Court records also indicate that some of these struggles were very drawn out, with some continuing for over fifty years with alternating actions of hedge

---

[114] Ibid 35–6.
[115] Ibid 37.
[116] Blomley, 'Making Private Property' (n 70) 5.
[117] Ibid 1–21, 5.
[118] Ibid 5; McDonagh (n 83) 37.
[119] McDonagh (n 83) 37.
[120] Lucy Finchett-Maddock, *Protest, Property and the Commons: Performances of Law and Resistance* (Routledge, 2016).

planting and breaking (sometimes accompanied by riots), the impounding of livestock and their rescue, and court actions from both sides.[121] The drawn-out nature of these conflicts helps to underscore the incremental and complex processes through which this new concept of private property gradually took root. In both the urban and rural contexts, most of the enclosures that were being protested were not themselves new – the hawthorn hedges that surrounded many outlying fields, for example, had been grown over many decades.[122] However, they had served different purposes – to 'mark parish boundaries' or as 'a defence [for crops] against cattle and sheep'.[123] Many of these could be described as temporary enclosures, which enabled 'private cultivation for part of the year' while allowing for common grazing for the remainder, and they were known as field closes.[124] What changed was the access being given to the enclosed pastures for grazing after the grain harvest – the 'half-yearly commoning rights' which usually took place 'from Lammas Day (1 August) or Michaelmas (29 September) to Candlemas (2 February) or Lady Day (25 March)'.[125] As land was converted to pasture, it became more profitable for private landholders to increase their own flocks of sheep and to attempt to keep their gates closed to the commoners' livestock.[126] The timing of many enclosure riots – to coincide with the beginning of these traditional commoning rights – emphasises this link.[127]

More's *Utopia*

It was into this political context, and the rising tensions around enclosure, that Thomas More, then Sheriff of London, published a seminal work of both fiction and socio-political satire that critiqued the times and portrayed a place (or rather 'no place') where enclosure was unthinkable and communal ownership of property the norm.[128] First published in December 1515 as *The best state of a Commonwealth and the new island of Utopia: A truly golden handbook, no less beneficial than entertaining, by the distinguished and eloquent*

---

[121] McDonagh (n 83) 38–9.
[122] Ibid 42–3; Liddy (n 83) 49.
[123] Blomley, 'Making Private Property' (n 70) 6.
[124] Liddy (n 83) 49–50; McDonagh (n 83) 42.
[125] Liddy (n 83) 49–50.
[126] McDonagh (n 83) 42–3; Liddy (n 83) 50.
[127] McDonagh (n 83) 42–3; Liddy (n 83) 50–1.
[128] Bruce Mazlish, 'A Tale of Two Enclosures: Self and Society as a Setting for Utopias' (2003) 20(1) *Theory, Culture & Society* 43, 45, citing Thomas More, *The Complete Works of St. Thomas More. Volume 4: Utopia*, ed. E Surtz, SJ and J H Hextor (Yale University Press, 1965) (*Utopia*, ed. Surtz and Hextor); B R Goodey, 'Mapping "Utopia": A Comment on the Geography of Sir Thomas More' (1970) 60(1) *Geographical Review* 15.

author, Thomas More, citizen and sheriff of the famous City of London,[129] ultimately, the title *Utopia* proved catchier.

More's *Utopia* is essentially a dialogue between the fictionalised narrator, Thomas More or 'Morus', the fictional Portuguese traveller, Raphael Hythlodaeus or 'Hythloday' and their mutual friend, Peter Giles (who was 'in reality the town clerk of Antwerp and a close friend of Erasmus' – a colleagues of More's[130]). In Book 1, Hythloday critiques English society and its entrenched inequality. Hythloday is particularly scathing about the English nobility, noting 'these evil men with insatiable greed have divided up among them all the goods which would have been enough for all people'.[131] In Book 2, Hythloday provides further context to his critique of sixteenth-century English society via a comparison with the democratic, essentially communist society of the island of Utopia (where he lived for five years before returning to Europe to share what he learned).

In Book 1, Morus and Giles try to convince Hythloday that he would make a good counsellor to a king. Hythloday explains why his counsel would be unwelcome and likely to incite ridicule, recounting by way of illustration a dinner conversation he once had in the late 1490s with Cardinal John Morton, Archbishop of Canterbury and Chancellor of England (and, in reality, More's former guardian). At the Cardinal's dinner table, a fellow guest (an English lawyer) was praising the execution of thieves and wondering aloud why so many remained at large despite the speed and number of executions being carried out. Hythloday responded by arguing that capital punishment was not only unjust but ineffective as a deterrent, since theft was the result of destitution rather than greed. When challenged, he went on to argue that poverty itself was the result of an unjust social order rather than laziness. As part of this analysis of the English social order, Hythloday segues into a strident (and now famous) critique of the practice of enclosure:

> [I]n whatever parts of the land the sheep yield the softest and most expensive wool, there the nobility and gentry, yes, and even some abbots though otherwise holy men, are not content with the old rents that the land yielded to their predecessors. Living in idleness and luxury, without doing any good to society, no longer satisfies them; they have to do positive evil. For they leave no land free for the plough: they enclose every acre for pasture; they

---

129 Wilde (n 69) 11.
130 Ibid 41.
131 More, *Utopia*, ed. Surtz and Hextor (n 128), cited in Timothy Kenyon, *Utopian Communism and Political Thought in Early Modern England* (Pinter Publishers, 1989) 68.

destroy houses and abolish towns, keeping only the churches, and those for sheep-barns. . . . Thus one greedy, insatiable glutton, a frightful plague to his native country, may enclose many thousand acres of land within a single hedge. The tenants are dismissed and compelled, by trickery or brute force or constant harassment, to sell their belongings. By hook or by crook these miserable people – men, women, husbands, wives, orphans, widows, parents with little children, whole families (poor but numerous, since farming requires many hands) – are forced to move out. They leave the only homes familiar to them, and they can find no place to go. Since they cannot afford to wait for a buyer, they sell for a pittance all their household goods.[132]

Hythloday's argument is mocked by those at the table, until Cardinal Morton gives it serious consideration, prompting his fellow diners to become suddenly receptive. Hythloday concludes, '[f]rom this episode you can see how little courtiers would value me or my advice.'[133] In response, with a line of argument that scholars believe reveals something of More's own motivations in writing *Utopia*,[134] Morus defends the project of seeking to sway the politics of the day by indirect means, by contending: 'If you cannot pluck up bad ideas by the root, or cure long-standing evils to your heart's content, you must not therefore abandon the commonwealth. Don't give up the ship in a storm because you cannot hold back the winds.'[135]

At the conclusion of Book 1, Hythloday finishes by arguing that he is

wholly convinced that unless private property is entirely abolished, there can be no fair or just distribution of goods, nor can the business of mortals be conducted happily. As long as private property remains, by far the largest part of the human race will be oppressed by a distressing and inescapable burden of poverty and anxieties.[136]

Nonetheless, Hythloday does grant that the introduction of specific reforms might help to alleviate the worst problems of poverty.[137] Lawrence Wilde

---

[132] Thomas More, *Utopia (Latin Text and English Translation)*, ed. George Logan, Robert Adams and Clarence Miller (Cambridge University Press, 1995) 63 (More, *Utopia*, ed. Logan et al.).

[133] Ibid 81.

[134] See, e.g., W B Gerard and Eric Sterling, 'Sir Thomas More's *Utopia* and the Transformation of England from Absolute Monarchy to Egalitarian Society' (2005) 8(1) *Contemporary Justice Review* 75, 87; Russell Osgood, 'Law in Sir Thomas More's Utopia as Compared to His Lord Chancellorship' (2006) 1 *Thomas More Studies* 177, 185–6.

[135] More, *Utopia*, ed. Logan et al. (n 132) 97.

[136] Ibid 103.

[137] Ibid 103–4.

argues that these 'intermediate measures' were particularly politically significant due to the inclusion of

> laws limiting the amount of land owned or income received, laws prevent-
> ing the prince from becoming too powerful or the populace too unruly, and
> a clampdown on corruption by making the sale of public offices illegal and
> ensuring that officials are properly remunerated.[138]

As discussed above, Parliament had already been attempting to limit
the practice of enclosure (and engrossment), and the new lord chancellor,
Cardinal Wolsey, was soon to appoint two commissions of Chancery to
investigate the practice further (in 1517 and 1518).[139] Wolsey's commissions
'led to a decree that all enclosures of the previous thirty-two years should
be abolished'.[140] However, although 'two hundred and sixty-four powerful
landowners were arraigned before the Court', these landowners 'were, more
often than not the same people who administered the law, and Wolsey's
good intentions were thwarted'.[141] The related issues of limiting monarchical
power and reducing corruption were further from the agenda, and went on
to 'become the central political issue in the next century, culminating in the
Civil Wars of the 1640s and the temporary abolition of the monarchy and
the House of Lords'.[142]

In Book 2, Hythloday describes the society and governance of Utopia in
some detail. The most significant features of Utopia are the complete rejection of private property, the egalitarian ethos, and the decentralised, democratic system of government. Utopians do not own houses; instead, they use
a lottery system to rotate dwellings every decade.[143] Hythloday argues that
these common property arrangements prevent the misappropriation of land,
such as those evidenced by the enclosure movement.[144] These arrangements
also help to protect social and economic equality, which is further enhanced
through the practice of shared labour.[145] Since everyone engages in labour
(either in the fields or in practical crafts) and no one is attempting to accu-
mulate more than they need, the work day in Utopia is just six hours long.[146]

---

[138] Wilde (n 69) 44, citing More, *Utopia*, ed. Surtz and Hextor (n 128).
[139] Wilde (n 69) 44.
[140] Ibid.
[141] Ibid.
[142] Ibid 44–5.
[143] Gerard and Sterling (n 134) 77.
[144] Ibid 78.
[145] Ibid.
[146] Ibid.

This frees up everyone's time to engage in intellectual pursuits and to socialise – a theory reflected in the Paris Commune's 'communal luxury' considered in Chapter 4. The resulting equality also flows into Utopia's system of government, under which power is bestowed by popular vote rather than resulting from entrenched hierarchy or the accumulation of wealth.[147]

Wilde argues that the 'elevation of labour as the basis of social freedom is a strong theme in Utopia and was to become a central tenet of all socialist thought'.[148] He also highlights the influence of Plato's *Republic*, 'where the ruling guardian class are denied possessions to ensure that they are unable to use public power for their own material benefit'.[149] In *Utopia*, More has extended this denial to all citizens, but the principle remains the same. Another similarity is the attempt in both the *Republic* and *Utopia* to synergise the public interest with private interests. As Hythloday puts it, 'where there is no private business, every man zealously pursues the public business'.[150]

Here we can see some of the intersecting debates around private property, popular sovereignty, liberty and social justice. On the one hand, private property is being critiqued for distorting democracy and social justice. On the other, the solution of collapsing the distinction between private and public interests (and lives) seems to imply a total lack of privacy and autonomy, which may have implications for liberty and personal sovereignty. (Private) property is, on the one hand, reified (particularly in the United States) as a bulwark against government overreach and a necessary precondition for personal liberty. However, property is conversely a communitarian compact that is a source of obligation, including compliance with societal norms and expectations. Furthermore, it was communal, not private, property that was the source of urban commoners' franchise or their stake in civic governance, as performed when 'riding the bounds'. In modern progressive theory, this historical example has echoes in the idea of property being an entrance to community, where ownership instils not only rights but obligations and expectations of civic participation.[151]

It is important to put these apparent compromises within their historical context. As Wilde points out, 'at this point in history the idea of individual liberty, in the liberal sense of freedom from external direction, was virtually unheard of'.[152] He concedes that '[m]ost modern readers would consider

---

[147] Ibid 85.
[148] Wilde (n 69) 48.
[149] Ibid 51.
[150] More, *Utopia*, ed. Surtz and Hextor (n 128) 103, cited in Wilde (n 69) 51.
[151] Eduardo M Peñalver, 'Property as Entrance' (2005) 91(8) *Virginia Law Review* 1889.
[152] Wilde (n 69) 52.

life [in Utopia] to be stultifying and joyless, and certain aspects of it would have seemed deeply unattractive to many sixteenth-century readers.'[153] But he argues that not only was More's Utopia taking care of everyone's material needs (in stark contrast to the endemic poverty in England), but it also enabled unprecedented levels of popular sovereignty.[154] If anyone was unhappy with the social constraints on liberty, '[t]he mechanisms for social change [were] in their hands.'[155]

More's decision to make his book a work of fiction, along with his use of irony and dialectic dialogue, has resulted in a somewhat ambiguous text that has confounded interpretation for centuries. George Logan blames the lack of consensus as to the meaning of More's *Utopia* on 'an excessive gap of sophistication between it and its readers'.[156] He acknowledges that '[t]he most insightful early readers of Utopia that we can identify are the group of More's fellow humanists represented in the commendatory letters prefixed to early editions.'[157] But he goes on to argue that 'these readers seem determined to find a book that embodies their own social and religious ideals'.[158] He criticises historians for ignoring More's extensive use of irony and 'treat[ing] *Utopia* as if it were a simple manifesto'.[159] Students of literature are equally targeted for 'the unpardonable sin of trivializing More's impassioned profoundly reflective and enormously learned book as a *jeu d'esprit*'.[160] Ultimately, Logan argues that 'the ironic fiction in which More embodies his views serves . . . to render both the full complexity of these views and the author's own complex attitude towards them'.[161]

Most scholars agree that More was clearly using the voice of Hythloday to critique the increasing disparity of wealth in early sixteenth-century England, and to safely satirise both the upper classes and capitalism more broadly.[162] Logan, for example, concludes that,

> however difficult it may be to determine what More thinks of Utopia, it is perfectly clear what he thinks of the present condition of England and Europe. . . . The indictment [of poverty, enclosure and the idleness of the

---

[153] Ibid 70.
[154] Ibid 70–1.
[155] Ibid.
[156] George Logan, *The Meaning of More's Utopia* (Princeton University Press, 1983) 3.
[157] Ibid.
[158] Ibid 4.
[159] Ibid.
[160] Ibid 5.
[161] Ibid 269.
[162] See, e.g., Gerard and Sterling (n 134) 71, 87; Wilde (n 69) 56–61.

upper classes] is important in itself . . . and, by underlining the gravity and urgency of European problems, it forms a perfect starting point for an exploration of the possibilities of reform.[163]

It is the meaning of Book 2 that elicits more debate. William Gerard and Eric Sterling argue that 'More believed so strongly in the ideals espoused in Utopia that he published them, hoping that he would influence his readers to reform English society.'[164] Others have pointed out that More's Utopian commons was not far removed from English customary law. Stephanie Elsky, for example, argues that More's discursive commons shares 'the same contours – and same contradictions – as English common law'.[165]

Despite this, the satirical dialectic dialogue between Hythloday and Morus creates ambiguity. While Hythloday's critiques of sixteenth-century English society and private property are sharp, Morus makes significant objections to alternatives suggested by the society of Utopia.[166] At the conclusion of Hythloday's long account of life in Utopia, Morus narrates:

> When Raphael had finished his story, many things came to my mind which seemed very absurdly established in the customs and laws of the people described – not only in their methods of waging war, their ceremonies and religion, as well as their other institutions, but most of all in . . . their common life and subsistence – without any exchange of money.[167]

For readers familiar with Greek, the satire was also apparent in the ironic meaning of the names given to Utopia's people and places.[168] In Greek, 'Utopia' means 'no place', 'Hythloday' means 'pedlar of nonsense', the Utopian capital 'Amaurot' means 'phantom', its governor, 'Ademis' means 'without a people', and its main river, the 'Anyder', means 'waterless'.[169] Some scholars have reasoned that these devices demonstrate that More ultimately saw his ideal society as unattainable or, at least, unsustainable. Another entirely plausible explanation is that he was merely being self-deprecating.

In contrast, Logan contends that 'More's Utopian construct embodies the results of a best-commonwealth exercise performed in strict accordance

---

[163] Logan (n 156) 49.

[164] Gerard and Sterling (n 134) 87. See also Wilde (n 69) 87.

[165] Stephanie Elsky, 'Common Law and the Commonplace in Thomas More's *Utopia*' (2013) 43 *English Literary Renaissance* 181, 184.

[166] Wilde (n 69) 11.

[167] More, *Utopia*, ed. Surtz and Hextor (n 128) 245, cited in Logan (n 156) 143.

[168] Wilde (n 69) 12.

[169] There is also some evidence to suggest that More originally intended to title the work 'Nusquama', which means 'nowhereland' in Latin, Wilde (n 69) 11.

with the Greek rules'.[170] Logan then highlights a number of 'evident defects' in Utopian society – including 'their method of waging war, their ceremonies and religion'. He concludes that rather than writing a pure vision of his ideal best-commonwealth, More's Utopia actually reflects a 'humanist realism' about the kind of Utopia that we could realistically hope to enact (or, at least, that could realistically be enacted in sixteenth-century Europe).[171] In support of this analysis, he argues:

> If More could have remedied the defects of Utopia (without violating the rules of the best-commonwealth exercise), presumably he would have: it is hard to imagine why he would play the best-commonwealth game without playing it as well as he could. The fact that he did not remedy these defects, together with the fact that he sometimes appears to go out of his way to call attention to them, suggests that he regarded them (like the discrepancies with Greek conclusions) as important findings – important results, that is, of the tests that he intended the exercise to perform.[172]

Wilde believes these rhetorical devices are deliberately ambiguous and unsettling, 'requiring the reader to think carefully about the issues raised'.[173] He argues that Utopia can be read as an argument for social justice, pointing out that contemporaneous 'commentators in More's lifetime regarded the book as an important social commentary, emphasising either its critique of existing institutions or its constructive ideas for new ones'.[174] Wilde concludes:

> With the advantage of hindsight we can see enclosure as an early example of the rise of production for profit at the expense of production to meet social needs, and indeed Marx quotes More's Utopia when making this point in the first volume of Capital.[175]

As discussed above in relation to the early enclosure riots:

> An additional and important consequence [of enclosure] was the shift of social power away from the poor, for the loss of the participatory management of the common field system deprived them of the most important source of collective power over their lives.[176]

---

[170] Logan (n 156) 139.
[171] Ibid 270.
[172] Ibid 243.
[173] Wilde (n 69) 11.
[174] Ibid 14.
[175] Ibid 38, citing Karl Marx, *Capital Volume I* trans. Ben Fowkes (Penguin, 1990) 880.
[176] Wilde (n 69) 58, citing J A Yelling, *Common Field and Enclosure in England, 1450–1850* (Macmillan, 1977) 215–16.

These radical political implications of More's *Utopia* provide some hint as to why the ironic, fictional approach may have been the safest option. As Wilde puts it, '[b]ehind the fictitious and often playful form of the book, and the objections provided by Morus, the author delivers a swingeing critique of the prevailing power structure in Europe.'[177] In doing so, he (like so many before him) also draws on religion to underscore the moral foundations of these critiques. Early in Book 1, for example, Hythloday points out that private property (and its social effects) goes against most of the teachings of Christ. This too would not have been without risk. As Wilde notes, 'More would have been mindful of the role played in the Peasants' Revolt in England in 1381 by the priest John Ball, who preached radical egalitarianism'.[178] The spectre of this past revolution and the fear of potential social collapse that might follow the growing calls for a new social order grounded in economic equality may help to explain the degree of attention More paid to the stringent regulation of public order in his fictional Utopia.[179]

It is clear from both the critique of English society and the structure of Utopia that More saw a strong and problematic connection between the accumulation of wealth and power, and social and spatial injustice. While Utopians take an extreme approach to reducing the perceived value of gold and precious gems (by using them to make chamber pots and children's playthings), the complete rejection of private property and the adoption of a comprehensive alternative common property regime are significant challenges to capitalism. Timothy Kenyon argues that 'More's "theory of property" is discernible only by inference and reconstruction.'[180] Nonetheless, by reading between the lines, he surmises that, while More appeared to believe that 'natural law is neutral as to the most appropriate form of ownership', inevitably private property will be used to ill effect due to the fallen nature of man.[181] More argued that 'wherever you have private property and money is the measure of all things, it is hardly ever possible for a commonwealth to by just or prosperous'.[182] Writing in 1888, Karl Kautsky described More as 'the first modern socialist',[183] while Gerard and Sterling argue that *Utopia* was a

---

[177] Wilde (n 69) 41.
[178] Ibid 60.
[179] Ibid 61.
[180] Kenyon (n 131) 71.
[181] Ibid 72.
[182] More, *Utopia*, ed. Logan et al. (n 132) 101.
[183] Karl Kautsky, *Thomas More and His Utopia* (Lawrence and Wishart, [1888] 1979) 94, cited in Gerard and Sterling (n 134) 14.

radical 'gesture of idealistic revolution that continues to reverberate in the 21st century'.[184]

Conclusion

This chapter considered a period characterised by population decline, recurrent waves of the plague, the withering of the manorial system, and an emerging political theory and identity around the commons and the commonweal. Through the demands of the Peasants' Revolt, the direct action of anti-enclosure rioters, and the literary arguments of More, we glimpse an emerging idea of *utopia* – an *other* way of relating to space and community. The next chapter picks up the story just as the population was beginning to recover and grow, and these new identities, politics and property practices were coming under increased pressure. During the early sixteenth century, the increasing commercialisation of agriculture and the related shift to a new proto-capitalist economy sparked new methods of resistance. If the period captured in this chapter provides us with utopian glimpses, the commoners of the next era filled in the blanks by putting these utopian ideals into concrete practice.

---

[184] Gerard and Sterling (n 134) 76.

4

A Concrete Utopia

OAK (ENGLISH)

Several decades after the first publication of *Utopia*, as the growing rural population was faced with a scarcity of land and paid labour, increasingly intense waves of anti-enclosure riots swept across rural England. Commoners uprooted hedges, levelled ditches, destroyed gates and rioted in the streets. There was also violence, though the gentry was responsible for the majority of this.[1]

This civil unrest, which was partly fuelled by popular resistance to the Henrician Reformation, culminated upon the death of Henry VIII in what is sometimes known as the 'Mid-Tudor Crisis'.[2] At the height of this crisis was a period known as the 'commotion time', when significant rebellions erupted in counties throughout England. The best known of these was Kett's Rebellion, during which thousands of rebels from all over Norfolk and Suffolk set up camp at Mousehold Heath outside Norwich – a camp that has been described as 'the greatest practical utopian project of Tudor England and the greatest anticapitalist rising in English history'.[3]

Significantly for this chapter, this rebel camp was the beginning of a largely new approach to popular resistance against enclosure; an approach under which demands for communal property rights and economic justice came to be pursued through the establishment of concrete utopias. Fleeting though they may have been, these utopias – including the rebel camp at Mousehold Heath, the Diggers colony on St George's Hill and the Paris Commune – all moved beyond theory into the physical world in order to manifest working models of an alternative social order or, less prosaically, to create liminal spaces of possibility.

This chapter starts with an examination of the commotion time,

---

[1] Jane Whittle, 'Lords and Tenants in Kett's Rebellion 1549' (2010) 207(1) *Past & Present* 3, 29 ('Lords and Tenants').

[2] Whitney R D Jones, *The Mid-Tudor Crisis, 1539–1563* (Macmillan, 1973).

[3] Jim Holstun, 'Utopia Pre-empted: Kett's Rebellion, Commoning, and the Hysterical Sublime' (2008) 16 *Historical Materialism* 3, 5.

including Kett's Rebellion and the reverberations that followed. We then move forward almost a century to the English Civil Wars and the establishment of the Diggers commune on St George's Hill. In this section, we consider the historical context in which the Diggers colony was established, as well as the political and religious manifestos of its leader, Gerrard Winstanley. Like Kett's Rebellion, the concrete utopia established by Winstanley and the Diggers was temporary in nature, but its impact lingered in the form of Winstanley's writing and in the ideals that were brought to life.

The third section of this chapter crosses the Channel to briefly explore the Paris Commune and the socialist utopia that was established in 1871. Though the Paris Commune can be distinguished from Kett's rebel camp and the Diggers colony in terms of both its urban context and its expanded scope, there were common threads, particularly in relation to its commitment to communal property relations. The Paris Commune can also be seen as a culmination of new ideals that were first tentatively explored at Mousehold Heath, before being given fuller scope by the Diggers. The first of these was the shift from bounded local commons to a universal claim to common property – an ideal that included demands for economic equality.

The second ideal, which is linked to the first, related to an increasingly strong commitment to popular sovereignty. While the rebels at Mousehold Heath and the Diggers corresponded with the state leaders – Lord Protector Edward Seymour in the first instance and Oliver Cromwell in the second – they also established their own parallel systems of governance. Although they did not directly challenge the monarchy, the rebels at Mousehold Heath asserted their right to be involved in questions of governance. By the time of the Diggers commune, broader political events had already overthrown the Crown, and Winstanley particularly defined his politics in opposition to 'Kingly rule'. But while Cromwell's new regime was re-establishing elite rule, Winstanley asserted a more egalitarian (albeit patriarchal) vision of popular sovereignty. Finally, the Paris Commune was firmly committed to the right of every person, regardless of gender or national origin, to participate in governance. Indeed, in many ways the Paris Commune sought to move beyond the state entirely, seeing it as inextricably tied to working-class oppression.

The chapter concludes by moving forward almost a century again to 1968, to consider the ongoing reverberations of the events of the nineteenth century and evolving claims to the right to the city.

The Commotion Time

In the period immediately following the death of Henry VIII, during the brief minority reign of Edward VI, England was governed by a regency council, which was initially headed by the king's uncle, Edward Seymour, 1st Duke of Somerset and known as Protector Somerset. In the fourteenth and fifteenth centuries, the labouring classes in England had experienced an extended period of improved living standards and increasing autonomy.[4] Indeed, Karl Marx described the period from 1300 to 1450 as the 'golden age of labour emancipating itself' in England.[5] However, by the mid-1500s, population growth, the ongoing commercialisation of agricultural, and an 'exceptionally bad harvest'[6] in 1545–6 all contributed to a downward pressure on wages and upward pressures on commodity and land prices.[7] The result was a dramatic increase in poverty and an expanding landless class.[8] Increasing competition over land also created new social tensions over access to the commons. When land had been abundant and cheap, a relatively relaxed approach had been taken to the use of the commons by all classes, including the gentry. With poverty, hunger and landlessness on the rise, these liberties were perceived in a different light and commoners began to push back against enclosure with new vigour.

This push back began in earnest in 1548, with a series of 'disturbances' across many parts of England.[9] These localised 'outbreaks of disorder' can be distinguished from the rebellions of the following year, but they also served as a prelude to the 1549 commotions in that they took place in similar parts of England and were focused on similar complaints – objections to the Reformation and economic grievances relating to enclosure, seigneurialism,

---

[4] Ibid 7.

[5] Karl Marx, *Grundrisse*, in 'Economic Manuscripts of 1857–61', in K Marx and F Engels, *Collected Works. Volume 28: Marx 1857–1861* (International Publishers, 1986) 433, cited in Holstun (n 3) fn 10.

[6] C S L Davies, 'Slavery and Protector Somerset: The Vagrancy Act of 1547' (1966) 19(3) *The Economic History Review* 533, 538.

[7] Andy Wood, *The 1549 Rebellions and the Making of Early Modern England* (Cambridge University Press, 2007) 30; Whittle, 'Lords and Tenants' (n 1) 40–1.

[8] Jane Whittle, *The Development of Agrarian Capitalism: Land and Labour in Norfolk 1440–1580* (Oxford University Press, 2000) 102, 107, 110, 152, 167, 175, 190.

[9] Amanda Jones, 'Commotion Time: The English Risings of 1549' (PhD thesis, University of Warwick, 2003) 3–4, citing John Hales, 'Defence', printed in E Lamond (ed.) *A Discourse on the Commonweal of This Realm of England* (Cambridge University Press, 1893) lviii; Wood (n 7) 40.

and the rising prices of land and food.[10] By the summer of 1549, the scale and intensity of the rebellion presented a serious challenge to the social and political order of the day.

Underscoring the significance of these events, historian Amanda Jones emphasises that '[t]he 1549 commotions were more than a series of enclosure riots; they broke out of the confines of the local community and became more generalised protests. In the eyes of the Tudor government, the "commotion time" constituted serious disorder.'[11] Similarly, in his contemporary critical correspondence to Protector Somerset, Sir William Paget warned that the commotions were 'a fundamental threat to the social order'.[12]

Historian Andy Wood quotes Antonio Gramsci to describe this period as a 'crisis of hegemony, or general crisis of the State'.[13] He argues that the crisis operated at multiple levels, in that it concerned economics, social relations, government policy, popular politics and religious belief.[14] Similarly, Jim Holstun argues that the rebellion 'grew out of long-term economic changes in agrarian relations, medium-term political precipitants, and a short-term cultural trigger'.[15] The politics of space and land use were central to most of these issues.

In June 1548, in response to increasing popular sensitivities over enclosure, Protector Somerset established a royal commission to investigate and report illegal enclosures.[16] This enclosure commission was headed by Somerset's ally John Hales, who, like Somerset, appears to have been on the side of the commoners and in favour of agrarian reform.[17] However, the remaining commissioners were selected from the gentry, and were widely perceived to have a vested interest in enabling the enclosure of the commons – a perception that is supported by evidence that they intimidated and threatened commoners seeking to give evidence.[18] As a result, rather than quelling popular unrest, it appears that the enclosure commissions simply added fuel to the fire and provided a new forum for the growing social conflict between the gentry and the commons.[19] Again, in his contemporary correspondence,

---

[10] Jones (n 9); Wood (n 7) 40.

[11] Jones (n 9) 10.

[12] Wood (n 7) 23.

[13] Ibid 21, citing Quintin Hoare and Geoffrey Nowell Smith (eds) *Selections from the Prison Notebooks of Antonio Gramsci* (Lawrence & Wishart, 1971) 210.

[14] Wood (n 7) 21.

[15] Holstun (n 3) 6.

[16] Wood (n 7) 38–9.

[17] Holstun (n 3) 7.

[18] Wood (n 7) 38–9.

[19] Ibid 39.

Paget warned Somerset that his 'social reforms, directed against enclosure and seigneurial oppression, had spun out of control, alienating the traditional ruling class and allowing the rebellious commons into government'.[20]

Paget also noted the role of the Henrician Reformation in creating the crisis of legitimacy.[21] However, the Reformation appears to have played a complex role in this civil unrest. On the one hand, some of the protests were partly a reaction to the Reformation, particularly the confiscation of ecclesiastical property and the destruction of symbolic artefacts. Religious institutions had long served as centres of education and community life, in addition to being a primary source of accommodation and alms for the poor. The rapid dissolution of monasteries by Henry VIII in the lead-up to Edward VI's reign had left significant gaps in the social safety net and resulted in core community resources being sold off. As a result, these aspects of the Reformation were deeply unpopular.[22]

On the other hand, religious reforms overturning existing religious tenets and institutions that had governed popular conduct for centuries had created the space for 'other fundamentals of the political system to be questioned by ordinary people'.[23] The Church had played a significant role in justifying and upholding the existing social and political order. By challenging the legitimacy of the Church, the Reformation inadvertently challenged the legitimacy of everything else, including social inequality.[24] Furthermore, by acknowledging the capacity of lay people to personally read and interpret the Scriptures, the Reformation implicitly acknowledged their capacity to play a role in governance – giving new life to ongoing calls for popular sovereignty.

As we saw in Chapter 3, *commonweal* ideology – championed in this era by the so-called 'commonwealth men' – was emphasised in anti-enclosure riots and other demonstrations, and appeared to hold a strong degree of legitimacy amongst the commons. Against this ideology, the behaviour of the gentry came under significant challenge as being both illegal and, perhaps more fundamentally, illegitimate. Wood reports that '[f]or a group of mid-sixteenth-century radical writers whom posterity has labelled the

[20] Ibid 22, citing John Strype (ed.) *Ecclesiastical memorials relating chiefly to religion and the reformation of it and the emergencies of the Church of England under King Henry VIII, King Edward VI and Queen Mary I*, 4 vols (Clarendon, 1822) 2: II, 429–37; for manuscript versions, see BL, Cotton MS, Titus 3 B, fols 277r–279v; PRO, SP10/8/4. See also his comments in PRO, SP68/4, fols 53r–54r (no. 185), fols 71r–72v (no. 189).

[21] Wood (n 7) 23.

[22] Ibid 24.

[23] Whittle, 'Lords and Tenants' (n 1).

[24] Wood (n 7) 23.

"commonwealth men", the answer to the crisis that had been opened up by the Reformation was obvious: it lay in a combination of preaching, teaching and social reform.'[25] These writers, who included Hugh Latimer, Robert Crowley, Henry Brinklow and John Hales (the man Protector Somerset appointed to head up his enclosure commission),[26] drew on the theological tradition of Wycliffe to critique the gaps in the social safety net created by the dissolution of monastic property and to call for a wider distribution of wealth.[27] Like More and the rebels of the commotion time, these writers were particularly critical of the gentry's participation in the commercial wool economy due to the way it fuelled enclosure and reduced access to paid labour.[28]

The rise of the printing press allowed these critiques to be widely distributed, including in the form of popularly accessible pamphlets that denounced the gentry for 'failing in their duties to the commons' and for opposing the enclosure commissions.[29] Crowley was particularly critical of the gentry's abandonment of the stewardship principles of property, arguing that 'commodity fetishism lay at the roots of social conflict'.[30] Similarly, Latimer asserted that 'no rich man can say before God, "This is my own." No, he is but an officer over it.'[31] He 'warned the gentry that their enclosures and depopulations would create an alienated, atomised society'.[32]

The disturbances

The first of the early disturbances took place in Cornwall on 5 April 1548, when a crowd rose up in response to government attempts to remove 'superstitious' images and murdered an unpopular government agent, William Body.[33] The riots then grew to some 3,000 people, before being repressed by an army levied by the local gentry.[34] In Hertfordshire, in early May, rumours that an enclosure commission was going to be held to allow Sir William Cavendish, lord of the manor of Northaw, to enclose part of Northaw Great Waste, led to large-scale enclosure riots and protests.[35] The trouble had been

---

[25] Ibid 30.
[26] Ibid 30–6.
[27] Ibid 31–3.
[28] Ibid 34.
[29] Ibid 34–5.
[30] Ibid 36.
[31] Ibid, citing G E Corrie (ed.) *Sermons of Hugh Latimer, sometime Bishop of Worcester, Martyr, 1555*, Parker Society, 22 (Cambridge, 1844) 411 (Latimer, *Sermons*).
[32] Wood (n 7) 36–7, citing Latimer, *Sermons* (n 31) 109.
[33] Wood (n 7) 40.
[34] Ibid.
[35] Jones (n 9) 33; Wood (n 7) 41.

brewing for some years, and culminated with a group of some 500–700 armed rioters besieging Cavendish's house, breaking down his hedges and fences, destroying his rabbit warrens, killing over a thousand rabbits, and threatening him with murder.[36] Armed rioters also confronted the enclosure commissioners on Northaw common in order to 'petitions to them in ther right for the comon',[37] causing the commissioners to flee without executing their commission.[38] Jones describes this incident as a key spark of the commotion time.[39]

In a theme that would repeat itself the following year, the organisers of the 1548 protests were enmeshed in the village power structures. In Hertfordshire, for example, '[t]he organisers of the protest included both the town constable of Cheshunt and the high constables of the hundred.'[40] It is also clear from accounts of the protests that participants felt justified in their actions and believed the law – or at least Protector Somerset and, thus, the king – was on their side. One man, for example, is recorded as remarking that protestors 'would defend their common & kepe it untyll the K[ing]s matie came to hys Full age'.[41] In Yarmouth, protestors emphasised the legitimacy of their complaints due to the impact of food price increases and the enclosure of the commons on the poor, and that they had been driven to action because the 'insasyable & gredy or covetous' town elite had worked to prevent 'the commons from articulating their complaints'.[42]

Kett's Rebellion

In April the following year, 'Somerset issued his second declaration against enclosures, empowering commissioners to rectify recent enclosures of common land.'[43] The declaration triggered 'widespread rioting, petitioning and demonstration', which was often met with severe repression from the local gentry.[44] By the summer of 1549, these brewing hostilities had reached new heights and resulted in the eruption of widespread civil unrest across most of England. This summer of rebellion included the East Anglian revolt, which comprised six rebel camps[45] and was characterised in the aftermath as

---

[36] Jones (n 9) 33; Wood (n 7) 41.
[37] Wood (n 7) 41, citing yeoman John Thompson.
[38] Jones (n 9) 34.
[39] Ibid 35.
[40] Wood (n 7) 41.
[41] Ibid.
[42] Ibid 43.
[43] Ibid 48.
[44] Ibid.
[45] Whittle, 'Lords and Tenants' (n 1) 10.

'the campyng tyme'.[46] The longest lasting and most infamous of these camps was established in July 1549 by somewhere between 6,000 and 20,000 rebels at Mousehold Heath, outside the city of Norwich.[47] It came to be known as Kett's Rebellion, after a yeoman named Robert Kett who took a leadership role in its establishment.

The specific events of the Norfolk rebellion can be traced back to Saturday, 6 July 1549, when crowds gathered at the site of Wymondham Abbey (which had recently been seized under Reformation policies) to watch the locally significant 'Wymondham game' – an illegal play commemorating St Thomas Becket's relics.[48] The play was followed by a traditional (and also illegal) feast, which wrapped up on Monday, 8 July.[49] By this stage, revellers were in high spirits and, 'inspired by the news that rebels in Kent had risen', decided to carry out a bit of levelling (the destruction of hedges and fences as a practical and symbolic act against enclosure).[50] One of their first targets was John Flowerdew, a minor gentleman in nearby Hethersett.[51] Flowerdew had a long-standing conflict with Robert Kett of Wymondham and paid the rioters to target Kett's property instead of his own.[52] However, something unusual occurred when the rioters arrived at Kett's estate: instead of defending his property, he welcomed their levelling efforts and offered to lead their rebellion in an attempt 'to subdue the power of Great men'.[53]

Wood describes Kett as 'an unlikely rebel'.[54] He was a wealthy landholder with 'a certain faculty for accumulating possessions'.[55] Like many established yeoman of the time, 'he had previously been prosecuted in the manor court for enclosing part of the town's commons'.[56] It is impossible to be certain of Kett's motivations for joining the rebellion. However, 'both he and his brother William (who joined him as a leader of the insurrection) had been members of the town guild of St Thomas Becket', and his long-standing conflict with

---

[46] Diarmaid MacCulloch, 'Kett's Rebellion in Context' (1979) 84 *Past & Present* 36, 38.

[47] Wood (n 7) 63. 'Kett's Rebellion' was infamous in that it was widely and vehemently condemned in the period following its demise. See, e.g., Alexander Neville, *Norfolkes Furies, or a View of Ketts Campe* trans. R Woods (2nd ed., 1615) STC.

[48] Jones (n 9) 10; Wood (n 7) 60.

[49] Jones (n 9) 10; Wood (n 7) 60.

[50] Wood (n 7) 60.

[51] Holstun (n 3) 7; Wood (n 7) 60.

[52] Wood (n 7) 60; Holstun (n 3) 7.

[53] Neville (n 47) sigs B3r–v.

[54] Wood (n 7) 61.

[55] Holstun (n 3) 8.

[56] Wood (n 7) 61.

Flowerdew partly stemmed from his 'opposition to Flowerdew's attempts to strip the assets of Wymondham Abbey following its dissolution'.[57] While it has been suggested that Kett acted primarily out of class resentment (due to his status just below the gentry), Holstun argues that 'this ignores the suicidal strangeness of his action'.[58] Instead, he speculates that Kett was motivated by a range of factors, including commonwealth ideology.[59]

It was Kett who led the rebels to Mousehold Heath, where they established camp on 11 July 1549.[60] In this 'highly-structured open-air commune',[61] an alternative government was created, 'made up of representatives of Norfolk and Suffolk hundreds, headed by Robert Kett'.[62] From the seat of an ancient oak tree, known as the Oak of Reformation, in Thorpe Wood, this rebel council issued warrants, and (after kidnapping the wanted gentlemen) held court and imprisoned those found guilty in the former palace of the Earl of Surrey, where Kett took up residence.[63]

The campers also ran a thriving camp kitchen – with food requisitioned from the local gentry – and held regular church services, using Cranmer's new prayer book.[64] Jane Whittle notes that '[s]heep flocks were particular targets, providing food and a symbolic attack on gentry property.'[65] The impact of this was significant, with some gentlemen recording a loss of up to 2,000 sheep during the rebellion.[66] There is also evidence that the campers were supported by crowd funding, with representatives from North Elmham and Carleton Colville in Suffolk, for example, being paid wages and supplied with food to join the rebellion.[67]

While the initial protests in Norwich in early July consisted primarily of locals, rebel campers who gathered at Mousehold Heath came from many localities across East Anglia, with those from eastern Norfolk being best represented in the records.[68] Whittle reports that '[s]ome communities seem

---

[57] Ibid.
[58] Holstun (n 3) 8.
[59] Ibid.
[60] Wood (n 7) 62–3; Holstun (n 3) 8–9.
[61] Holstun (n 3) 9.
[62] Wood (n 7) 62–3.
[63] Ibid 63.
[64] Whittle, 'Lords and Tenants' (n 1) 19; Holstun (n 3) 9.
[65] Whittle, 'Lords and Tenants' (n 1) 19, citing Aubrey Greenwood, 'A Study of the Rebel Petitions of 1549' (PhD thesis, University of Manchester, 1990) 268–9.
[66] Whittle, 'Lords and Tenants' (n 1) 19, citing Aubrey Greenwood, 'A Study of the Rebel Petitions of 1549' (PhD thesis, University of Manchester, 1990) 268–9.
[67] Holstun (n 3) 9; Whittle, 'Lords and Tenants' (n 1) 11.
[68] Whittle, 'Lords and Tenants' (n 1) 12–13.

to have joined the rebellion *en masse*.'[69] For example, manorial courts records indicate that twenty-six tenants from Blickling, 'including most of the manorial jury, were ordered to make new fealty to their lord in autumn 1549 because they had gone "to Mousehold"'.[70] Overall, the rebels were broadly representative of the general population, with records indicating that many of the leaders had previously been involved in a dispute either over common rights or over the preservation of Wymondham Abbey.[71] Given that many viewed the confiscation of Church property as a form of enclosure, it could be argued that resistance to enclosure was the most common thread – and this interpretation is supported by the contents of the petitions drafted by the rebels.

Somerset corresponded with a number of rebel leaders during the commotion time, including those at Mousehold Heath, mostly by way of royal herald.[72] Wood argues that '[t]hese letters highlight the confused nature of Somerset's response to the commotion time, combining standard authoritarian denunciations of plebeian rebellion with sometimes wild offers of compromise.'[73] Through this correspondence, the rebels at Mousehold Heath petitioned the king, calling, amongst others things, for a reduction of rents and food prices, and that 'no man shall enclose any more'; 'lord of no manor shall common upon the common'; 'rivers may be free and common to all men for fishing and passage'; and 'all bond men may be made free, for God made all free with his precious bloodshedding'.[74] They also demanded that 'the commons should select Somerset's enclosure commissioners', due to the popular belief that many of the existing commissioners were corrupt.[75]

The Mousehold petition was evidently written by committee – it contains a number of particularly specific demands, a high degree of repetition (particularly on the issue of access to the commons), and was signed by representatives of thirty-three hundreds, reflecting the distributed leadership structure at the camp.[76] Whittle argues that, '[l]ike other aspects of the rebellion, the structure of hundred representatives mimicked official power

---

[69] Ibid 11.
[70] Ibid, citing Blickling manor court roll, NRO, NRS 11265 26A5, court held the day after St Michael the Archangel (September 1549).
[71] Whittle, 'Lords and Tenants' (n 1) 23.
[72] Wood (n 7) 64.
[73] Ibid 53.
[74] Excerpted in Frederic William Russell, *Kett's Rebellion in Norfolk* (Longmans, 1859) 48–56.
[75] Ibid 55.
[76] Whittle, 'Lords and Tenants' (n 1) 37.

structures within the county, where two high constables were chosen for each hundred to report crimes and misdemeanors to the quarter sessions.'[77] This broad representation amongst the leadership reflects the historical context from which the rebellion emerged, and Whittle argues that previous disputes had served to clarify particular grievances related to enclosure and engrossment and had 'created local leaders'.[78] She explains, that '[w]hen legal cases were brought by communities against a manorial lord, representatives had to be chosen, legal advice sought, and a common purse collected to finance the case: this experience could easily feed into rebellion.'[79]

While the Mousehold petition contained only one article on enclosure (and it was tempered by a locally specific call for saffron crops to be exempted), direct action against enclosure both kicked off the rebellion and continued throughout.[80] Records indicate that, in addition to camping, many rebels 'took action closer to home, putting down enclosures and carrying out various partly symbolic acts'.[81] The official anti-enclosure position, coupled with the recent establishment of the enclosure commission, supported the general sentiment that enclosures lacked legitimacy – meaning that levelling could be justified.[82] The case record from the Star Chamber in the lead-up to the rebellion (and in the immediate aftermath) indicates that the biggest impact of enclosures was on access to the commons – an issue that is addressed repeatedly in the petition.[83]

On 21 July 1549, a royal herald offered a pardon to the rebels at Mousehold Heath. The pardon was accepted by some, but Kett argued that he was not in arms against the Crown and, thus, had no need of pardon.[84] From this point on, the rebels were officially classified as traitors, and the city of Norwich was closed to them.[85] Less than a week later, Northampton sent around 1,500 men to Norfolk. Wood surmises that this 'small force was probably intended to persuade the rebels to submit'.[86] Instead, they attacked and 'Lord Sheffield, Northampton's second-in-command, was killed.'[87] The

---

[77] Ibid 38. Whittle notes that '[h]undreds were also a unit for tax collection', fn 152.
[78] Ibid 39.
[79] Ibid.
[80] Ibid 42.
[81] Ibid 8.
[82] Ibid 43.
[83] Ibid, citing Greenwood (n 65) 254, 258–9.
[84] Wood (n 7) 64.
[85] Ibid.
[86] Ibid 65.
[87] Ibid 66.

remaining troops fled the city and the rebels were able to establish control over Norwich.[88]

By early August, the regency council had decided serious military force was necessary to quash the rebellion, and raised an army of 'at least 6,000 foot soldiers and 1,500 calvary' comprising both gentry and foreign mercenaries.[89] Somerset was initially set to lead the force himself, but ultimately the Earl of Warwick was appointed as commander.[90] Negotiations were attempted but were unsuccessful and the royal army was vicious in its response. In one example, Warwick 'captured forty-nine rebels whom he promptly hanged at the market cross'.[91] Fighting continued for a number of days, eventually concluding on 28 August 1549. By this time, the rebels had rejected three offers of royal pardon, which required them to disperse without their demands being met.

In the final battle of Kett's Rebellion, the rebels were soundly defeated by the royal army. Between 2,000 and 3,500 rebels are estimated to have died in battle,[92] and a further 250 royal soldiers.[93] In the aftermath, many rebels were arrested for their involvement, including Robert and William Kett (who both fled as the final battle turned to catastrophe).[94] In his contemporary, unsympathetic account of the events, Alexander Neville documents that judgement began immediately after the final battle, 'and many were hanged and suffered grievous death. Afterward, nine which were the ringleaders and principals, were hanged on the oak called the Oak of Reformation.'[95]

These executions continued for over a week, with the final number of executions estimated to reach around 300.[96] Robert and William Kett were captured and taken to the Tower of London for trial. After being found guilty of treason, there were sentenced to be hanged – Robert from the walls of Norwich Castle and William from the west tower of Wymondham Abbey, the site of the festival that had set the Norfolk rebellion in motion.[97] Wood concludes, 'Kett's defeat left the rebellious commons traumatized and broken.'[98]

---

[88] Ibid.
[89] Ibid 67.
[90] Ibid.
[91] Ibid.
[92] Ibid 68; Whittle, 'Lords and Tenants' (n 1) 21.
[93] Whittle, 'Lords and Tenants' (n 1) 20.
[94] Ibid; Wood (n 7) 68.
[95] Neville (n 47), cited in Whittle, 'Lords and Tenants' (n 1) 20.
[96] Wood (n 7) 71–2.
[97] Ibid 73.
[98] Ibid 69.

Reverberations of the Rebellion

Wood argues that '[t]he intense violence that marked Kett's rebellion is difficult to explain without some understanding of the particular sharpness of social conflict in early and mid-sixteenth-century Norfolk.'[99] He emphases the 'peculiar rapaciousness of the gentry',[100] which followed familiar patterns of increasing rents, overstocking and exploiting the commons, and other forms of blatant appropriation and disrespect of customary law,[101] while also including brutal attempts to 'reimpose serfdom'.[102] Wood summarises, '[w]hen this seigneurial offensive was combined with the assertiveness and independence of the commons of Norfolk, who were used to defending their rights through organised collective litigation, petition, demonstration and riot, the result was explosive.'[103]

Other historians have adopted different interpretations of the prominence of Kett's Rebellion. Jones, for example, questions the uniqueness of the Norfolk rebellion, suggesting that similar events in Hertfordshire and Middlesex were equally significant, despite being less well documented.[104] Whittle also challenges Wood's emphasis on 'the peculiar rapaciousness of the gentry' in Norfolk,[105] noting the evidence that the same underlying causes of lower-class anger were present across England. Wood may be correct that the Norfolk gentry were particularly extreme in their attempts to 'reimpose serfdom', but the rising landless class elsewhere was still highly vulnerable to the newly passed Vagrancy Act of 1547, which provided that 'vagabonds' could be enslaved for two years.[106] It also provided that the master of such vagabond slaves could put rings of iron on their neck and legs, feed them on bread and water, and 'cawse the saide Slave to worke by beating, cheyninge or otherwise in such worke and Labor how vyle so ever it be'.[107]

The rebellion appears to have been started by smallholders and the growing landless class,[108] who formed a pragmatic alliance with members of the rising yeoman class – wealthy landowners, who had been busy engross-

[99] Ibid 55.
[100] Ibid.
[101] Ibid 55, 57–8.
[102] Ibid 56.
[103] Ibid 57.
[104] Jones (n 9).
[105] Whittle, 'Lords and Tenants' (n 1) 47.
[106] Davies (n 6) 534.
[107] Ibid.
[108] Wood (n 7) 173–4; Whittle, 'Lords and Tenants' (n 1) 40–1.

ing their increasingly large holdings and profiting from the economic situation of the time.[109] It is not clear what motivated those in the lower gentry to join the rebellion.[110] It may have been that they were competing directly with the gentry and thus particularly sensitive to their abuse of the commons and the restrictions imposed by the existing social hierarchy.[111] Or they may have been motivated by Reformation theology or *commonweal* ideology.

The Mousehold petition largely reflected the demands of the wealthier rebels,[112] and formed part of a larger conversation they were engaged in with the government, especially Protector Somerset, who was clearly committed to some form of agrarian reform.[113] Whittle argues that the 'Mousehold petition was a tool for negotiation, not a comprehensive list of grievances, and it is the interests of the poor that were not articulated'.[114] Of course, it is also the case that the wealthier rebels were more focused on this process of negotiation, while the smallholders and the landless favoured direct action.[115]

State repression followed the defeat of the rebellion. Wood reports that, '[b]y the autumn of 1549, the market crosses and gates of towns across much of England were decked out with such grotesque reminders of the commotion time.'[116] In October 1549, Protector Somerset was arrested by the regency council and replaced with John Dudley, Earl of Warwick – the man who had led the final battle against Kett's Rebellion. Popular dissent, including enclosure rioting and protest, continued, despite the swift repression that generally followed.[117] This included a significant enclosure riot in Northaw in 1579, which involved many of the same individuals or families as those involved in the riots of 1548 (and the riots that preceded these in 1544),[118] and the Oxfordshire Rising of 1596.[119] Jones argues that the heavy-handed state response to these risings indicates that the regime had learned

---

[109] Wood (n 7) 173–4; Whittle, 'Lords and Tenants' (n 1) 5.

[110] Whittle, 'Lords and Tenants' (n 1) 25–6.

[111] Ibid 5.

[112] Wood (n 7) 163, 172.

[113] Whittle, 'Lords and Tenants' (n 1) 45, citing Ethan H Shagan, *Popular Politics and the English Reformation* (Cambridge University Press, 2003).

[114] Whittle, 'Lords and Tenants' (n 1) 45.

[115] Ibid 46.

[116] Wood (n 7) 76.

[117] Jones (n 9) 333–45; Wood (n 7) 76.

[118] Jones (n 9) 334, 338.

[119] Ibid 340–2.

their lesson from the commotion time and were paranoid about the risk of large-scale popular disorder.[120]

Holstun contends that the rebellion was part of a drawn-out clash between two distinct 'visions of culture and society'.[121] The rebel campers used 'commoning rhetoric and practice, . . . to try to restore the moral economy of the country community',[122] and to 'use the resources of a newly centralized Tudor state to preserve and extend the independence enjoyed by English small producers during the fifteenth century'.[123] In contrast, 'Tudor gentlemen . . . [used] pre-emptive decisionist violence, to crush the Norwich commune, overthrow Somerset, and accelerate capitalist primitive accumulation'.[124]

The broad demographic representation amongst participants in the rebellion, and the length of time over which popular dissent continued, indicates widespread dissatisfaction with the status quo and a particular consensus around the fundamental illegitimacy of elite actions to overstock and enclose the commons. The violent repression of the rebellion – which continued well into the second half of the sixteenth century – also indicates that the introduction of private property and capitalist modes of production was made possible only through the use of brute elite power. It was neither inevitable nor popular.

The Second Revolution

While the role of the Reformation in the commotions was complex, the influence of theology in fuelling popular dissent increased during the late sixteenth century due to the proliferation of radical Protestant sects, whose preachers roamed the countryside spreading the ideas that came to feed a revolution. The Anabaptists were one such sect, and Christopher Hill reports that their 'name came to be used in a general pejorative sense to describe those who were believed to oppose the existing social and political order'.[125] As 'William Gouge told his shocked City congregation in the 1620s, . . . [Anabaptists] teach that all are alike and that there is no difference betwixt masters and servants.'[126] Another sect was the Familists, 'members of the Family of Love',[127] who 'held their property in common, believed that all

---

[120] Ibid 344.
[121] Holstun (n 3) 3.
[122] Ibid.
[123] Ibid 6.
[124] Ibid 3.
[125] Christopher Hill, *The World Turned Upside Down: Radical Ideas During the English Revolution* (Penguin Books, 2019) 13 (*World Turned Upside Down*).
[126] Ibid 19–20, citing W Gouge, *Of Domesticall Duties* (1626) 331–2.
[127] Hill, *World Turned Upside Down* (n 125) 13.

things come by nature, and that only the spirit of God within the believer can properly understand the Scripture'.[128]

While this unrest settled somewhat during the Elizabethan era, it rose again with the coronation of her cousin, James I (James VI of Scotland) and reached fever pitch during the reign of his son Charles I. Existing 'class antagonism was exacerbated by the financial hardships of the years from 1620 to 1650', which ultimately resulted in the English Revolution.[129] However, while the Revolution was a reaction to class hostility, its actors did not take a unified approach. Instead, Hill argues there were essentially 'two revolutions in mid-seventeenth century England'.[130] The first, successful, revolution partly mirrored the Magna Carta in establishing political and property rights for the elite, including the 'abolition of feudal tenures', protection against arbitrary taxation, guarantee of the 'sovereignty of Parliament and common law', and the 'abolition of prerogative courts'.[131] It also helped to pave the way for a new economic system by elevating the 'protestant ethic',[132] which Max Weber argues is the foundational theory of capitalism.[133]

The second, unsuccessful, revolution was more akin to the Forest Charter of 1217, in that it sought to establish 'communal property, a far wider democracy in political and legal institutions', and went even further in terms of rejecting both the state Church and the Protestant ethic.[134] It is this second, commoners' revolution that interested Hill because of the significance of the ideas being promoted and the fact that, for a time at least, they seemed capable of succeeding. Key groups within this second revolution included political groups like the Levellers and Diggers; radical Protestant sects like the Baptists, Quakers and Muggletonians; and 'sceptical' groups like the Seekers and Ranters.[135] It is important to note that many of these groups overlapped, with all of them raising sceptical questions about existing social institutions and beliefs, and with religion informing almost all of their radical approaches.[136] Indeed, it was this religious underpinning that drove much of the fervour of the commoners' revolution, with a royalist comment-

---

[128] Ibid.

[129] Ibid 8.

[130] Ibid 3.

[131] Ibid. These were achieved by way of the Tenures Abolition Act 1660 (sometimes known as the Statute of Tenures) and the Statute of Frauds of 1677.

[132] Hill, *World Turned Upside Down* (n 125) 3.

[133] Max Weber, *The Protestant Ethic and the Spirit of Capitalism: And Other Writings* (1905) trans. Peter Baehr and Gordon C Wells (Penguin Books, 2002).

[134] Hill, *World Turned Upside Down* (n 125) 3.

[135] Ibid 2.

[136] Ibid.

ing in 1648, '[a]ll sorts of people dreamed of an utopia and infinite liberty, especially in matters of religion.'[137] Hill describes this time as:

> a period of glorious flux and intellectual excitement, when, as Gerrard Winstanley put it, 'the old world . . . is running up like parchment in the fire.' Literally anything seemed possible; not only were the values of the old hierarchical society called in question but also the new values, the protestant ethic itself.[138]

It is fascinating to consider what kind of society might have been created in England if this second revolution had succeeded. Nonetheless, beyond this thought experiment, the potency and contours of the central tenets of this thwarted revolution remain relevant to us today. They represent the shadow doctrines of another system of law in relation to property, rights and governance that continue to resonate, and continue to bubble to the surface in times of conflict and flux. The ongoing potency of these shadow doctrines also seems to indicate that they carry a level of legitimacy that elite rule can never claim.

Winstanley and the True Levellers (Diggers)

Winstanley associated enclosure with man's 'fallen' nature and the Norman Conquest.[139] Timothy Kenyon argues that Winstanley initially viewed enclosure and the establishment of a system of hired labour, along with man's capacity to dominate the environment, as all being aspects of 'the curse' that occurred as a consequence of the Fall.[140] However, he came to revise his views on dominion, arguing that it could be justified through labour – specifically digging:[141] 'Winstanley employed the theory of the Norman Yoke to exemplify the worst aspects of the human condition . . . [He] regarded Normanism or "kingly power" as the most abject social manifestation of fallen human nature.'[142] Though the historical accuracy

---

[137] W Chestlin, *Persecutio Undecima* ([1648] 1681) 8, cited in Hill, *World Turned Upside Down* (n 125) 19.

[138] Hill, *World Turned Upside Down* (n 125) 2–3.

[139] See, e.g., ibid 106; Timothy Kenyon, *Utopian Communism and Political Thought in Early Modern England* (Pinter Publishers, 1989) 121–52.

[140] Kenyon (n 139) 131, citing Gerrard Winstanley, *Several Pieces Gathered into One Volume* (Manchester Public Library) 155 (*Several Pieces*).

[141] Kenyon (n 139) 132, citing Gerrard Winstanley, *The Works of Gerrard Winstanley*, ed. G H Sabine (Cornell University Press, 1941; reprinted Russell & Russell, 1965) 564 (*Works*), cross-referencing Gerrard Winstanley, *Winstanley: The Law of Freedom and Other Writings*, ed. C Hill (Penguin, 1973; Cambridge University Press, 1983) (*Law of Freedom*) 347–8.

[142] Kenyon (n 139) 132, 141.

of 'the Norman Yoke' is open to question, Kenyon argues that 'the vision of a "classless" Anglo-Saxon society, supposedly destroyed at the historical watershed of 1066, yet partially recovered through concessions wrested in the Magna Carta, was deeply imbued in the consciousness of Englishmen up to and beyond the seventeenth century'.[143] This was particularly prevalent during the English Civil War, with many radicals, 'particularly the Levellers, . . . [deploying] the theory of the Norman Yoke alongside their conceptions of natural rights'.[144]

Winstanley argued that since the Civil War had been fought to over-throw Normanism, it required the restoration of Man's birthright to cultivate the land (at least with respect to the wastes and commons).[145] Specifically, he argued, '"[t]hat government that gives liberty to the gentry to have all the earth, and shuts out the poor commons from enjoying any part, . . . is the government of imaginary, self-seeking Antichrist," and must be rooted out.'[146] While the political leaders of the Levellers were focused on power, Winstanley was more concerned with economic inequality and the spiritual restoration of man. In aid of both of these concerns, Winstanley advocated for economic communism – 'a society in which "no man shal have more land, then he can labour himself, or have others to labour with him in love, working together, and eating bread together"'.[147]

Winstanley commenced digging on St George's Hill on 1 April 1649 in the unsettled period after the execution of Charles I.[148] Here Winstanley sought to claim the commons 'for and in behalf of all the poor oppressed people of England and the whole world' and, through digging, to create a new society free from exploitation.[149] He named his community 'the True Levellers', but they came to be called 'the Diggers' and this was the name that stuck. While some have explained the Diggers commune as a consequence of scarcity and necessity, Kenyon argues that '[d]igging was squatting of a unique nature in which the occupation of the wastes and commons was

---

[143] Ibid 141, citing Christopher Hill, 'The Norman Yoke', in C Hill, *Puritanism and Revolution: Studies in Interpretation of the English Revolution of the 17th Century* (Secker & Warburg, 1958) 51.

[144] Kenyon (n 139) 142.

[145] Ibid 132, 133.

[146] Hill, *World Turned Upside Down* (n 125) 108, citing Gerrard Winstanley, *The Works of Gerrard Winstanley*, ed. G H Sabine (Cornell University Press, 1941) 385, 395, 472.

[147] Kenyon (n 139) 144, 166, citing Winstanley, *Several Pieces* (n 140) 191.

[148] Kenyon (n 139) 168.

[149] Ben Maddison, 'Radical Commons Discourse and the Challenges of Colonialism' (2010) 108 *Radical History Review* 29, 35, citing Winstanley, *Law of Freedom* (n 141) 109.

justified by arguments that are central to an intricate social theory.'[150] He also points out that 'Winstanley did not emphasize the argument from necessity until well into the Digger period, at which point the colony was actually threatened by extinction.'[151]

Painting a picture similar to the conditions that led to the commotion time, Keith Thomas describes the Diggers movement as 'the culmination of a century of unauthorized encroachment upon the forests and wastes by squatters and local commoners, pushed on by land shortage and pressure of population'.[152] The Diggers believed the land they occupied was commons, wastes and Crown land,[153] and argued that it had 'returned again to the Common people of England' upon the king's execution – a novel attempt to transfigure the legal forest back into a lawful forest.[154] Echoing claims made in the lead-up to the Forest Charter of 1217, and in More's *Utopia*, Winstanley argued that this common 'land had been stolen from the people and "hedged into Inclosures" by the rich while the poor lived in "miserable poverty"'.[155] Grounding his arguments in natural law, Winstanley asserted that the earth was made by the Lord, to be a common Treasury for all, not a particular Treasury for some.[156]

The Civil War period was marked by an unusually free press and was a time when an abundance of radical ideas were printed and circulated often in the form of political pamphlets. In 1640, as Charles I's reign entered the political crisis of the Long Parliament, pre-publication censorship broke down and there was an immediate proliferation of printed material.[157] John

---

[150] Kenyon (n 139) 169.

[151] Ibid.

[152] Keith Thomas, 'Another Digger Broadside' (1969) 42 *Past and Present* 57, 58, cited in Hill, *World Turned Upside Down* (n 125) 92.

[153] Gerald Aylmer, 'The Diggers in Their Own Time', in A Bradstock (ed.) *Winstanley and the Diggers, 1649–1999* (Routledge, 2013) 8, 16–17; Briony McDonagh and Carl J Griffin, 'Occupy! Historical Geographies of Property, Protest and the Commons, 1500–1850' (2016) 53 *Journal of Historical Geography* 1, 5.

[154] *The Speeches of the Lord General Fairfax and the Other Officers of the Armie, to the Diggers at St George's Hill in Surrey and the Diggers Several Answers and Replies Thereunto* (Printed for RW, 1649), quoted in McDonagh and Griffin (n 153) 5.

[155] McDonagh and Griffin (n 153) 5, citing Gerrard Winstanley, 'A Declaration to the Powers of England', in T N Corns, A Hughes and D Loewenstein (eds) *The Complete Works of Gerrard Winstanley*, vol. 2 (Oxford University Press, 2009) 5.

[156] Gerrard Winstanley, *The True Levellers Standard Advanced: Or, The State of Community Opened, and Presented to the Sons of Men* (1649), available at <https://www.marxists.org/re ference/archive/winstanley/1649/levellers-standard.htm>.

[157] John Rees, *The Leveller Revolution: Radical Political Organisation in England, 1640–1650* (Verso, 2017) 67–9.

Rees reports that '[p]rinted titles of all kinds rose from 600 to 700 a year in the 1630s, to 900 in 1640, to more than 2,000 in 1641, and to over 3,500 in 1642.'[158]

In the lead-up to the commotion time, in the mid-sixteenth century, the articulation of unsanctioned religious ideas had been identified as a serious threat to the social order. Wood notes, 'the preface to the 1549 edition of the English Bible condemned those who "*by theyr inordinate reading, undiscrete speaking, contencious disputing, or otherwise by theyr licentous lyvinge*" presented the Scriptures as a subject for debate'.[159] However, during that era, these ideas were mostly shared at 'the plebeian alehouse'[160] or by the occasional rebel clergy. By the mid-seventeenth century, the printing press had provided a new medium for radical ideas to be widely disseminated, and Rees argues that it was 'the Levellers' ability to define the political tasks in print and to use printed material to organise their followers that made them play a significant part in shaping the eventual outcome of the English Revolution'.[161] Winstanley cannot claim such political influence, but his writing – seven pamphlets and two longer works – did play its own role in shaping debate during the Civil War period and, more significantly, the fact that it was printed meant that his ideas continued to resonate long after the revolution was over and the monarchy restored.

In his first pamphlet, *An Appeal to the House of Commons* (11 July 1649), Winstanley seeks to convince the House of Commons that the Civil War had created a contract between Parliament and the people 'to restore the conditions that had existed prior to the Conquest and before the Fall'.[162] In his third pamphlet, *A New-yeers Gift for the Parliament and Armie* (1 January 1650), Winstanley ascribes the institutions of 'buying and selling', freeholding and copyholding to the Norman Conquest, along with the functions of the lords of the manor, the clergy and lawyers.[163] Parliament had already passed laws to abolish the monarchy and to establish the Commonwealth, and Winstanley believed that it was obliged to go further in order to fully

---

[158] Ibid 69.

[159] Wood (n 7) 25, citing the Bible in English (1549).

[160] Wood (n 7) 25.

[161] Rees (n 157) 76.

[162] Kenyon (n 139) 175–6, citing Gerrard Winstanley, *An Appeal to the House of Commons* (11 July 1649), in Winstanley, *Works* (n 141) 305 (*Appeal to the House of Commons*), cross-referencing Winstanley, *Law of Freedom* (n 141) 116.

[163] Kenyon (n 139) 179, citing Gerrard Winstanley, *A New-yeers Gift for the Parliament and Armie* (1 January 1650), in Winstanley, *Works* (n 141) 387 (Winstanley, *Law of Freedom* (n 141) 199–200).

eradicate the institutional remnants of the 'Norman Yoke'.[164] Given the pre-vailing theories of the time – which justified the Civil War and the English Revolution via the theory of the 'Norman Yoke' and appeals to natural rights (the *commonweal*) – Winstanley's framing of the inherent righteousness of the Diggers' claims was well placed to grant them a sense of legitimacy.

The Diggers focused on occupying only the commons, and thus seemingly presented no threat to existing enclosed property holdings. As Winstanley wrote, '[l]et the Gentry have their inclosures free from all Norman enslaving intanglements whatsoever, and let the common people have their Commons and waste lands set free to them, from all Norman enslaving Lords of Manors . . .'.[165] However, it is important to recognise that by this stage Winstanley had already laid out a vision for 'the subversive withdrawal of hired labour and the collapse of landlordism such action might precipitate'.[166] It may simply have been strategic not to highlight this aspect of his vision during the Diggers' period of occupation.

Winstanley's sixth pamphlet, *A Vindication of Those, Whose Endeavors is Only to Make the Earth a Common Treasury, Called Diggers* (4 March 1650) was published after the Diggers colony had been subjected to both legal and physical harassment, and eviction from St George's Hill, which required them to relocate to nearby Cobham.[167] Ultimately, this Diggers colony also failed and was disbanded. Nonetheless, Winstanley held on to his 'conviction that the spiritual and moral welfare of men could be attained only' via a com-munal lifestyle.[168] While ruling-class interests, and particularly a local gentle-man, Mr Willard Drake, did orchestrate the eviction of the Diggers, the involvement of local commoners highlights the clash between the Diggers' ideals of universal rights to the commons and a bounded approach based on local custom. Nonetheless, the Diggers' vision for a new society, a utopia, lived on, partly due to Winstanley's writing.[169]

Indeed, while Winstanley's colony on St George's Hill is the best known, Hill describes his political pamphlets as bearing 'fruit'.[170] As many as thirty-three or thirty-four other Diggers colonies were established all over southern

[164] Gerrard Winstanley, *A New-yeers Gift for the Parliament and Armie* (1 January 1650) (Early English Books Online Text Creation Partnership, 2011), available at <http://name.umdl .umich.edu/A96697.0001.001>.

[165] Winstanley, *Appeal to the House of Commons* (n 162) 306 (Winstanley, *Law of Freedom* (n 141) 116–17), cited in Kenyon (n 139) 176.

[166] Kenyon (n 139) 169.

[167] McDonagh and Griffin (n 153) 6; Maddison (n 149) 37; Kenyon (n 139) 169.

[168] Kenyon (n 139) 185.

[169] Maddison (n 149) 29–30.

[170] Hill, *World Turned Upside Down* (n 125) 89.

and central England, 'at Wellingborough in Northamptonshire, Cox Hall in Kent, Iver in Buckinghamshire, Barnet in Hertfordshire, Enfield in Middlesex, Dunstable in Bedfordshire, Bosworth in Leicestershire, and [other locations] in Gloucestershire and Nottinghamshire'.[171] Several of these colonies published their own pamphlets and many later became Quaker centres (Winstanley himself became a Quaker shortly after the demise of his Diggers colony).[172] Hill reports that these colonies 'had some influence in intensifying ill-feeling between landlords and tenants, . . . [and] may have contributed to the class consciousness of Fifth Monarchists and early Quakers'.[173]

In 1652, Winstanley published his final work, *The Law of Freedom*, which sets out his detailed vision for a communist system of society – a utopian manifesto.[174] In order to anticipate certain stock objections to communism, Winstanley emphasised that everyone would be required to work and notes 'exploitation, not labour, is the curse'.[175] Indeed, communal cultivation is a fundamental aspect of Winstanley's utopia, both for reasons of production and for its crucial role in the spiritual restoration of the individual.[176] Winstanley's proposed system for achieving this universal labour is fairly detailed: '"Non-productive" elements of contemporary society, such as lawyers, were excluded from Winstanley's utopia on the grounds of their social disutility.'[177] He stipulated 'that every young person . . . would be instructed in a trade or artifice' via a seven-year apprenticeship (already standard at the time), and 'patriarchal masters were expected to keep their households hard at work'.[178] That being said, he did advocate for 'retirement from physical labour at the age of forty'.[179] At this 'age of maturity', men would be ready to govern others, or to enter the clergy should they so wish to.[180]

And, yes, he did mean just men. Winstanley took a very conservative approach to the position of women in his utopian vision, arguing for a strong patriarchal role for husbands and fathers, and only limited education

---

[171] Ibid 89–91.

[172] Ibid 90.

[173] Ibid 91.

[174] Kenyon (n 139) 192.

[175] Ibid 195–6, citing Christopher Hill, 'Introduction', in Winstanley, *Law of Freedom* (n 141) 53.

[176] Kenyon (n 139) 197, citing Winstanley, *Works* (n 141) 577, 579, cross-referencing Winstanley, *Law of Freedom* (n 141) 362, 365.

[177] Kenyon (n 139) 196.

[178] Ibid.

[179] Ibid.

[180] Ibid 196, 204, citing Winstanley, *Works* (n 141).

for girls.[181] At least with respect to education, this contrasts with More's approach from over a century earlier, and likely reflects the decline in educational standard for women since the spread of Protestantism.[182]

Like More, Winstanley believed that poverty was the result of inequitable distribution (of both goods and labour) and the private property system.[183] To achieve equitable distribution, he proposed a system of storehouses. The first, 'general storehouses', would hold raw commodities that each family could collect for its trade. The second, 'particular storehouses', would hold the surplus goods manufactured by each of the trades and should be available to meet the needs of each family.[184] Crucially, there was to be no commodity market; no buying and selling of either goods or labour. He did appear to leave some room for some form of private property to the family patriarch in his house, furniture and 'anything which he hath fetched in from Storehouses, or provided for the necessary use of his Family'.[185] He also argued for some kind of ownership over 'sufficient working tools for common use',[186] but these appear to be held on trust for the common use of the community at large.

Briony McDonagh and Carl Griffin argue that by not focusing on specific parcels of land, Winstanley and 'the Diggers offered a much more radical critique of existing concepts of property and property rights as they were practiced in mid seventeenth century England than had the participants in anti-enclosure riots a century earlier'.[187] Ben Maddison explains that by engaging in such a critique, the Diggers and Winstanley reconceptualised the commons in three important ways.[188] First, they universalised the claim to the commons, shifting it from a locally bounded, exclusionary approach to one that was open to all. Second, they conceptualised this universal approach as a human right. And, finally, they reframed the idea of freedom as one that was both 'collective and defined by the ability to maintain independence from wage labor and commodity relations more generally'.[189]

---

[181] Kenyon (n 139) 212–13, citing Winstanley, *Works* (n 141) 95, 108.

[182] Kenyon (n 139) 213.

[183] Ibid 198.

[184] Ibid 199, citing Winstanley, *Law of Freedom* (n 149) 583–5.

[185] Kenyon (n 139) 201, citing Winstanley, *Works* (n 141) 512, cross-referencing Winstanley, *Law of Freedom* (n 141) 288.

[186] Kenyon (n 139) 199, citing Winstanley, *Works* (n 141) 550.

[187] McDonagh and Griffin (n 153) 5.

[188] Cristy Clark and John Page, 'Of Protest, the Commons, and Customary Public Rights: An Ancient Tale of the Lawful Forest' (2019) 42(1) *UNSW Law Journal* 26, 43, citing Maddison (n 149) 34.

[189] Maddison (n 149) 34.

By universalising the claim to the commons, 'Winstanley liberated commons conceptually from their everyday, practical enmeshments in the real property and power relations of seventeenth-century England.'[190] This universalisation of the commons – beyond the boundaries of the local inhabitants – was disputed by many at the time. The implication that 'the wastes were not privately owned was [considered by many to be] a wilful misunderstanding of the legal basis of common land which went against centuries of established legal and customary practice'.[191] Indeed, Winstanley's vision – especially in his essentially communist understanding of collective freedom and the need to resist the alienation of labour from the workers – challenged both the established social order, with its feudal traditions and customs, and the emerging capitalist society that was rapidly taking its place.[192] The language of rights, used by the Diggers when they referred to the commons as their 'creation birthright', also represented a radical conceptual shift and a challenge to the legitimacy of local custom that sought to exclude the landless from access to the commons.[193]

Background to the Paris Commune

The historical record of the two centuries that followed the English Civil War is full of stories of conflict, revolt and revolution. While issues of religion, personal liberty and popular sovereignty all featured heavily in these events, many were also bound up with questions of spatial justice. The passage of years saw the ongoing commercialisation of agriculture and the related enclosure of the commons. Dispossessed of their communal property rights, a growing landless class was forced into urban centres to serve as labour in the industrialising economy.

This shift to urbanisation and industrialisation represented a significant break from the past. Up until this point, the commons had formed the backbone of both the economic and political power of the labouring classes. These sites of communal property rights – this *lawful forest* – provided the raw material for subsistence living (firewood, honey, charcoal, agistment, pannage and common grazing land) and a crucial space for communal gatherings. Beyond this, the customary rights associated with the lawful forest were also the foundation of the stake in governance held by the commons, and of their demands for the *commonweal*. Forced off their lands into the city, a growing urban underclass had to locate a new basis for their ongoing demands for

---

[190] Ibid 35.
[191] Clark and Page (n 188) 43, citing McDonagh and Griffin (n 153) 5.
[192] Clark and Page (n 188) 43.
[193] Ibid.

economic survival, spatial justice and popular sovereignty. It is with this shift that claims to the *lawful forest* morphed into claims for the *right to the city*.[194]

Without their relational rights to communal property to ground their claims to power, working-class people turned to each other, including through the establishment of the labour movement. In 1836, for example (in the aftermath of the 1832 Reform Act, which had increased the suffrage from 10 per cent to just 18 per cent of the adult male population, thus thwarting working-class hopes for a voice in government[195]), the London Working Men's Association was established with the goal of seeking 'by every legal means to place all classes of society in possession of their equal, political, and social rights'.[196] The minute book of the Association from 18 October 1836 contains demands for 'Universal Suffrage, the protection of the Ballot, Annual Parliaments, Equal representation, and no property qualification for members'[197] – five of the six demands that were published two years later in *The People's Charter* (which was co-authored by William Lovett, one of the founders of the Association).[198]

Advocates of the Charter, known as 'Chartists', presented a petition with 1,280,958 signatures to Parliament in May 1839, but Parliament voted not to consider it.[199] Nonetheless, the campaign continued for two decades following its publication. The most significant event of the Chartist movement took place in April 1848 with a protest of some 100,000–200,000 people at Kennington Common, after which organisers presented Parliament with another petition (also rejected).[200] The timing of this protest coincided with a period of intense revolutionary activity that took place across Europe, which became known as the 'Printemps des Peuples' (Springtime of the Peoples).[201]

---

[194] See Henri Lefebvre, *The Right to the City* trans. Eleonore Kofman and Elizabeth Lebas [*Le Droit à la ville*, 1968] (The Anarchist Library, 1996); David Harvey, *Rebel Cities: From the Right to the City to the Urban Revolution* (Verso, 2012).

[195] David Avery, 'Chartism', British Library (15 May 2014), available at <https://www.bl.uk /romantics-and-victorians/articles/chartism>.

[196] *Minute Book of the London Working Men's Association*, British Library (18 October 1836), available at <https://www.bl.uk/collection-items/minute-book#>.

[197] Ibid.

[198] *The People's Charter*, British Library (c.1838), available at <https://www.bl.uk/collection -items/people-charter>. The sixth demand was that MPs be paid an annual salary of £500, so that working people could afford to serve in Parliament.

[199] Ibid.

[200] Lindsey German and John Rees, *A People's History of London* (Verso, 2012) 111; Avery (n 195).

[201] Eric Hobsbawm, *Age of Capital: 1848–1875* (Orion, 2010).

The spring of 1848 saw uprisings of millions of people 'from the Atlantic to Ukraine, from the Baltic to the Mediterranean'.[202] While these multiple revolutions had a wide variety of demands, they shared a focus on popular sovereignty, particularly including demands for an expansion of the franchise.[203] In France, where just 1 per cent of the population had a right to vote (with voting rights being linked to property ownership), the revolution began with a 'banquet campaign', which started in July 1847 with a series of private political meetings where speakers debated political reforms, including the expansion of the franchise, republicanism, and improving the living conditions of the working classes.[204] The banquets started out fairly moderate in tone, but became increasingly radical as time passed.[205] In a speech on 28 December 1847, King Louis-Philippe declared his opposition to any reform and this ultimately led to a vote the following February banning any further banquets.[206] While organisers voted to accept the ban,[207] the popular reaction to the forced cancellation of a banquet scheduled for 22 February was less sanguine, and the unrest that followed – during which the army refused to fight and the National Guard sided with insurgents – led the king to abdicate.[208]

In the wake of the king's abdication, a popular direct vote for government was held under the radical conditions of universal male suffrage.[209] However, with insufficient time for opposition parties or the population to prepare for the elections, established elites secured the overwhelming majority of votes.[210] This included Louis-Napoleon, who was elected President of the Second Republic.[211] By the end of 1851, Napoleon III had declared himself emperor and France was ruled by a monarch once more. Although Napoleon III, like many contemporaneous European governments, did introduce a series of reforms in response to the more moderate demands of the 1848 movement,

---

[202] Jonathan Sperber, *The European Revolutions, 1848–1851* (Cambridge University Press, 2nd ed., 2005).

[203] Ibid.

[204] John J Baughman, 'The French Banquet Campaign of 1847–48' (1959) 31(1) *The Journal of Modern History* 1–15.

[205] Ibid 1–11.

[206] Ibid 12–14.

[207] Ibid 14.

[208] Ibid; Sperber (n 202).

[209] Malcolm Crook, 'Universal Suffrage as Counter-Revolution? Electoral Mobilisation under the Second Republic in France, 1848–1851' (2015) 28(1) *Journal of Historical Sociology* 49.

[210] Ibid.

[211] Ibid.

the 1850s also saw a period of repression across Europe, due to 'concerted efforts to erase the memory of insurrection from public awareness'.[212]

Across the Channel, Karl Marx saw out the 1850s quietly as an exile in London, following the failed revolution in Germany.[213] However, after the dust had settled, the working-class movement began again to consolidate. On 28 September 1864, at St Martin's Hall in Long Acre, London, a meeting of 'English and French socialists and radicals, including Marx and Engels', established the International Workingmen's Association, which came to be known as the First International ('the International').[214]

When Napoleon III declared war on Prussia on 19 July 1870,[215] the International came out against the war, with the Paris section declaring it unjust and 'merely dynastic', and German branches declaring themselves 'the enemies of all wars, but above all dynastic wars', and 'happy to grasp the fraternal hand stretched out to us by the workmen of France'.[216] When Prussia defeated Napoleon III in early September 1870, occupied much of France and laid siege to Paris, the International came out in support of the people of France, while questioning the democratic credentials of the newly declared provisional Government of National Defence (GND).[217]

On 31 October 1870, after learning the GND were prepared to negotiate with Prussia, the workers of Paris and sections of the National Guard rose up and seized the Hôtel de Ville (City Hall), but were pacified the following day with the false promise of free elections.[218] On 22 January 1871, workers and members of the National Guard rose up again, only to be shot down by GND forces.[219] A week later, the siege of Paris was lifted as the city was surrendered to Prussian forces.[220] The National Guard were permitted to remain armed.[221]

---

[212] Christopher Clark, 'After 1848: The European Revolution in Government' (2012) 22 *Transactions of the RHS* 171, 171.

[213] German and Rees (n 200) 113.

[214] Ibid 119.

[215] Mitchell Abidor, *Communards: The Story of the Paris Commune of 1871, as Told by Those Who Fought for It* (Marxists Internet Archive Publications, Kindle ed., 2010) loc. 109.

[216] Cited in Karl Marx, 'The First Address' (23 July 1870), in K Marx, *The Civil War in France* (20th anniversary ed., 1891) 7 (*Civil War in France*).

[217] Karl Marx, 'The Second Address' (9 September 1870), in Marx, *Civil War in France* (n 216) 11–12; Abidor (n 215) loc. 144.

[218] Frederick Engels, '1891 Introduction', in Marx, *Civil War in France* (n 216) 1, 3 ('Introduction'); Abidor (n 215) loc. 151.

[219] Abidor (n 215) loc. 157.

[220] Engels, 'Introduction' (n 218) 3.

[221] Ibid.

On 26 February, Adolphe Thiers and Jules Favre, chief executive and member of the newly elected National Assembly, signed a preliminary peace treaty with Bismarck. By early March, the National Guard had begun to organise, leading Thiers to withdraw his government to Versailles and, on 18 March 1871, to attempt to disarm the National Guard.[222] To quote Frederick Engels, '[t]he attempt failed; Paris mobilized as one man in defence of the guns, and war between Paris and the French government sitting in Versailles was declared.'[223]

The Paris Commune

From this declaration of war on 18 March to their bloody defeat on 28 May 1871, the workers of Paris used their newly claimed autonomy to transform the city 'into an autonomous Commune whose social life was recalibrated according to principles of cooperation and association'.[224] In claiming their common rights to a stake in the governance of the city, the Communards made reference to the ancient providence of these rights: 'Paris aspires only to found the Republic and to restore its *communal franchises*, happy to provide an example to the other communes of France.'[225] Henri Lefebvre argues that '[t]he expression "franchises" precisely denotes the communal tradition under the Ancien Régime, with the extension to the urban bourgeoisie of the freedoms regained from the conquering race, the Franks.'[226]

Lefebvre goes on to note that the Commune's ideology was 'a very confused ideological-political complex, in which different and even contradictory aspects converge[d] and mingle[d]'.[227] This included the ideological approaches of the Proudhonists and the Blanquists, and 'the ideas that animated the International, which [were] themselves very complex'.[228] However, he argues, '[i]t was also utopia, insofar as they dreamed of a new life, established from one day to the next, born ardent and pure in the fire of

---

[222] Ibid 4; Abidor (n 215) loc. 175. For a contemporary, detailed account of the Versailles government's perspective on this event, see William Pembroke Fetridge, *The Rise and Fall of the Paris Commune in 1871* (Harper & Brothers, 1871) 28–32.

[223] Engels, 'Introduction' (n 218) 4.

[224] Kristin Ross, *Communal Luxury: The Political Imaginary of the Paris Commune* (Verso, 2015) 5.

[225] Henri Lefebvre, *The Proclamation of the Commune 26th March 1871* trans. David Fernbach (Verso, [1965] 2019), citing *The Paris Commune Manifesto* (6 April 1871) (emphasis added) (*Proclamation of the Commune*).

[226] Ibid.

[227] Lefebvre, *Proclamation of the Commune* (n 225).

[228] Ibid.

the Communal revolution and realising communitarian aspirations from one day to the next.'[229]

After so many years of thwarted activism and lengthy theoretical debates, the short life of the newly declared Paris Commune was a veritable flurry of activity to enact and experiment with a diverse array of concrete political and social reforms. The Commune acted immediately to deepen democracy and increase economic equality, by filling 'all posts – administrative, judicial, and educational – by election on the basis of universal suffrage of all concerned, with the right of the same electors to recall their delegate at any time',[230] and by declaring that all payments of rent for dwelling houses from October 1870 until April 1871 be remitted.[231] They also declared that the highest salary received by any member of the Commune, including the newly elected council members, could not exceed 6,000 francs.[232] Reflecting its socialist roots, the Commune announced 'the postponement of all debt obligations for three years and abolition of interest on them', began the process of organising workers into co-operative societies to take over manufacturing, and developed longer-term plans for the creation of one 'great union' for all workers.[233] Finally, turning their attention to issues of personal liberty, the Commune decreed religion to be a purely private matter, thus separating the Church from the state, and removed religious symbols and instruction from all schools.[234]

Of particular significance was the heavy involvement of both women and foreigners, with international solidarity – developed in good part through the International – and feminism both having a significance influence. Communards spoke of their flag as being 'the flag of the Universal Republic',[235] a concept that had its roots in a brief moment of internationalism during the 1789 revolution and had become popular again amongst members of the International. In addition to symbolising working-class solidarity, it was anti-imperialist in ambition and focused on dismantling state bureaucracy, particularly the standing army and police.[236]

---

[229] Ibid.
[230] Frederick Engels, '1891 Postscript', in Marx, *Civil War in France* (n 216) 40, 42 ('Postscript').
[231] Engels, 'Introduction' (n 218) 4.
[232] Ibid.
[233] Ibid.
[234] Ibid.
[235] *Journal Officiel de la république française sous la commune* [1871] (Éditions Ressouvenances, 1995) 103 (*Journal Officiel*), cited in Ross (n 224) 21.
[236] Ross (n 224) 22.

Kristin Ross describes the Commune as 'a working laboratory of political inventions, improvised on the spot or hobbled together out of past scenarios and phrases, reconfigured as need be, and fed by desires awaked in the popular reunions at the end of the Empire'.[237] Like others, she also emphasises the non-nationalistic originality of the Commune: 'It did not wish to be a state, but rather an element, a unit in a federation of communes that was ultimately international in scale.'[238] Given this anti-statist approach, scholars have argued that the Commune's origins are most accurately found in the clubs and reunions that began to meet in the dying days of the Second Empire, rather than in the actions of the state in attempting to seize the arms of the National Guard: 'For it was the reunions and the clubs that created and instilled the idea – well before the fact – of a social commune.'[239] This was also recognised by both the Communards and their opposition at the time, with anti-Communard Chevalier d'Alix defining the clubs and reunions as 'the Collège de France of insurrection'.[240]

Reminiscent of the private banquets that preceded the 1848 revolution, the people of Paris gathered in clubs and 'reunions' to listen to people debate political reforms. However, while these reunions included veterans of 1848, both the speakers and the audiences were drawn from far more diverse groups than the banquets, including 'young workers from the Paris section of the International Workers' Association and . . . refugees from London, Brussels, and Geneva'.[241] Ross reports that '[b]efore September 4, certain topics were policed and subject to censor', but this 'censorship of topics related to politics and religion had the paradoxical effect of enabling vaster, more imaginative speculation to take place'.[242] 'One could not speak against the emperor or his various functionaries but one could advocate for an end to private property.'[243] Also discussed, as early as 1869, were plans and practicalities for 'the Commune', with 'vive la Commune' being called out at the opening and closing of 'the more revolutionary clubs in the north of Paris'.[244]

The origins of the Commune could arguably also be traced to the Paris section of the International, which boasted around 50,000 members by 1870

---

[237] Ibid 11.
[238] Ibid 12.
[239] See ibid 14.
[240] Chevalier d'Alix, *Dictionnaire de la Commune et des communeux* (A Thoreux, 1871) 16.
[241] Ross (n 224) 15.
[242] Ibid 18.
[243] Ibid, citing *Les Orateurs des réunions publiques de Paris en 1869. Compte rendu des séances publiques* (Imprimerie Town et Vossen, 1869) 38.
[244] Ross (n 224) 19.

and was made up of a federated structure of 'local, independent worker-based committees' – a structure that was ultimately adopted by the Commune.[245] The pre-existence of all these things – the International, the National Guard, and the revolutionary clubs and reunions – all of which were interlinked and federated, is why many argue that the Paris Commune did not come into being on 18 March 1871; *'the Commune already in fact existed'.*[246] As Ross argues, 'the Commune was both rallying cry and the thing itself. Attempting to differentiate the two or establish the moment when the one was transformed into the other may be beside the point.'[247] Or, as Maurice Dommanget put it, '[t]he Commune was at the same time the thing and the rallying word, the reality and the sign, the fact and the ideology.'[248]

On 11 April 1871, the Women's Union was co-founded by twenty-year-old Elisabeth Dmitrieff, a founding member of the Russian section of the International, who had arrived in Paris from London on 28 March as 'a special correspondent sent by Marx to report back about the Commune for the International'.[249] The Women's Union sought to 'achieve the creation of a new social order of equality, solidarity and of freedom' and it did this by establishing co-operative workshops for women, seeking equal pay for women, and providing practical support for the Commune.[250] Ross argues that the Women's Union was 'the Commune's largest and most effective organization' and 'can be seen as the practical response to many of the questions and problems regarding women's labor that had been the discussion topic at the earliest popular reunions of 1868'.[251]

As part of its work to dismantle the bureaucracy of the state, the Commune sought to liberate both education and the arts from state control. In the case of education, this involved the establishment of free, compulsory, secular and, most importantly, holistic public education for all children.[252] The focus on holistic education was an attempt to 'overcome the division between manual and intellectual labor' and to claim for all children 'a right to

---

[245] Ibid.

[246] Arthur Arnould, *Histoire populaire et parlementaire de la Commune de Paris* (Librairie socialiste de Henri Kistemaeckers, 1978) 80, cited in Ross (n 224) 21 (emphasis in Arnould's original).

[247] Ross (n 224) 20.

[248] Maurice Dommanget, *Hommes et choses de la Commune* (Éditions École émancipée, 1937), cited in Lefebvre, *Proclamation of the Commune* (n 225).

[249] Ross (n 224) 20, 22, 24.

[250] John Milner, *Art, War and Revolution in France, 1870–1871: Myth, Reportage and Reality* (Yale University Press, 2000) 148.

[251] Ross (n 224) 27.

[252] Ibid 40.

intellectual life'.[253] On 15 May, a project was also proposed for the establishment of system of crèches, to be available throughout working-class areas[254]
– a project, of course, that never had the chance to be realised during the
short life of the Commune.

In the case of the arts, a 'Call to Artists' was issued in April for a political
meeting of Parisian artists (broadly defined).[255] Here Eugène Pottier read
out a Manifesto for an Artists' Federation, which called for 'the free expression of art, released from all government supervision and all privilege'.[256] It
concluded: 'We will work cooperatively toward our regeneration, the birth
of communal luxury, future splendors and the Universal Republic.'[257] Ross
argues that this call for *communal luxury* was a 'demand that beauty flourish
in spaces shared in common and not just in special privatized preserves'.[258]
In reference to the destruction of the Victory Column at Place Vendôme
(a potent symbol of Napoleon's Empire),[259] she goes on to suggest that 'we
might think of the demolition of the column as an initial clearing of the
terrain for communal luxury'.[260]

The Commune's emphasis on communal luxury and on everyone
working and, thus, everyone gaining time to engage in both artistic and
intellectual pursuits is reminiscent of both More's *Utopia* and Winstanley's
utopian manifesto, *The Law of Freedom*. In describing the 'true secret' of the
Commune, Marx argued that '[i]t was essentially a working class government,
the product of the struggle of the producing against the appropriating class,
the political form at last discovered under which to work out the economical
emancipation of labor.'[261] Emphasising the link between communal property
and self-government that comes up again and again over the course of this
book, Marx went on to argue that the economic emancipation of labour, via
the abolition of class property, was a crucial precondition to 'the Communal
Constitution', because '[t]he political rule of the producer cannot co-exist
with the perpetuation of his social slavery':[262]

[253] Ibid 42.
[254] Ibid 41.
[255] Milner (n 250) 148.
[256] *Manifesto of the Artists' Federation of Paris*, 15 April 1871, in *Journal Officiel, tome 2* (n 235)
273–4, cited in Ross (n 224) 50–1.
[257] *Manifesto of the Artists' Federation of Paris*, 15 April 1871, in *Journal Officiel, tome 2* (n 235)
273–4, cited in Ross (n 224) 39, 58.
[258] Ross (n 224) 58.
[259] See Fetridge (n 222) 265–9 for a detailed account of the destruction of the Victory
Column.
[260] Ross (n 224) 59.
[261] Marx, *Civil War in France* (n 216) 26.
[262] Ibid.

It is a strange fact. In spite of all the tall talk and all the immense literature, for the last 60 years, about emancipation of labor, no sooner do the working men anywhere take the subject into their own hands with a will, than uprises at once all the apologetic phraseology of the mouthpieces of present society with its two poles of capital and wages-slavery . . . The Commune, they exclaim, intends to abolish property, the basis of all civilization! Yes, gentlemen, the Commune intended to abolish that class property which makes the labor of the many the wealth of the few.[263]

The Versailles-based government made a point of highlighting the role of foreigners in the leadership of the Commune, thus 'othering' them sufficiently to justify their ultimately brutal treatment.[264] Indeed, members of the Paris section of the International had already recognised this tendency for the dominant class to frame the working-class as second-class citizens whose membership was contingent on their compliance with the established order.[265] Their international solidarity was explicitly anti-colonial as a result – particularly in relation to the French colonial occupation of Algeria.[266] In the aftermath of the Commune's demise, Communard Benoît Malon argued that the deadly 'ferocity' of the French military in reclaiming government control over Paris had been developed during its forty-year repression of Algeria.[267]

On 21 May 1871, Versailles troops entered Paris, and over the course of 'the bloody week' that followed, around 20,000–30,000 Parisians were killed.[268] Thousands more were summarily executed and buried in mass graves in the immediate aftermath.[269] Close to 45,000 were arrested, and while many were later released, some were executed and thousands were imprisoned or deported (most to New Caledonia).[270] Others managed to escape in the chaos of the battle and lived as exiles for many years . Many of these exiles found their way to London, where they remained politically active and were supported by Marx and Engels.[271]

---

[263] Ibid 26–7.

[264] Ross (n 224) 31.

[265] Ibid 31–2.

[266] Ibid 32.

[267] Benoît Malon, *La Troisième Défaite du prolétariat français* (G Guillaume Fils, 1871) 485–6, cited in Ross (n 224) 33.

[268] Engels, 'Introduction' (n 218) 5; Abidor (n 215) loc. 229–36; German and Rees (n 200) 125.

[269] Engels, 'Introduction' (n 218) 5; Abidor (n 215) loc. 236.

[270] Abidor (n 215) loc. 236.

[271] German and Rees (n 200) 126–7.

Though they were defeated in the end, the Communards fought back against the Versailles troops. During the drawn-out battle, Communards set fire to many of Paris's landmark buildings, including the Hôtel de Ville, the Palais des Tuileries, and even the Louvre library containing 100,000 volumes.[272] These acts of destruction, in addition to the conflict that existed within the Commune – particularly towards the end over whether to kill their hostages[273] – and the bitter recriminations that followed its demise,[274] made it easy to demonise the Commune (just as Kett's Rebellion before it had also been vilified).[275] This demonisation, and the state repression that followed,[276] has been blamed for the ultimate dissolution of the First International (though irreconcilable philosophical differences between anarchists and Marxists did not help).[277]

Certainly, the Paris Commune was far from perfect. Nonetheless, as Marx declared in the immediate aftermath, the most significant achievement of the Paris Commune was its 'own working existence'.[278] The Commune moved theory into practice and created a working model of an *other* way – a governance system that was shaped by workers from below, rather than being decreed by the elite from above. And it did this through a universalist, solidarity-based approach that rejected state power. As Engels argues:

> From the outset the Commune was compelled to recognize that the working class, once come to power, could not manage with the old state machine; that in order not to lose again its only just conquered supremacy, this working class must, on the one hand, do away with all the old repressive machinery previously used against it itself, and, on the other, safeguard itself against its own deputies and officials, by declaring them all, without exception, subject to recall at any moment.[279]

---

[272] Milner (n 250) 166–9; Abidor (n 215) loc. 5234. For a full 'List of all the public buildings, monuments, churches and houses, damaged or destroyed by the Communists of Paris' (as well as accompanying maps of the destruction), see John Leighton, *Paris under the Commune* (1871) 644–71.

[273] Leighton (n 272) 615–17.

[274] Eugene Vermersch, 'A Word to the Public', in Abidor (n 215) loc. 5410–40.

[275] See, e.g., Fetridge (n 222).

[276] See, e.g., Milner (n 250) 184–8.

[277] Lefebvre, *Proclamation of the Commune* (n 225); German and Rees (n 200) 126; Mikhail Bakunin, 'On the International Workingmen's Association and Karl Marx' (1872), in S Dolgoff (ed.) *Bakunin on Anarchy* (Vintage Books, 1971).

[278] Marx, *Civil War in France* (n 216) 29.

[279] Engels, 'Postscript' (n 230) 42.

David Harvey argues that the Paris Commune was both an anti-capitalist class struggle and 'an urban social movement that was reclaiming citizenship rights and the right to the city'.[280] Highlighting the spatial dimension of historic working-class struggles, including the Paris Commune, he notes that 'the dynamics of class exploitation are not confined to the workplace'.[281] Here he highlights the multiple practices of accumulation by dispossession by landlords and financiers that operate to extract wealth from the working class just as effectively as employers.[282] Significantly, Harvey also notes that 'urbanization is itself produced'.[283] Or, as Lefebvre puts it, the 'city is itself *oeuvre* [a body of work; a work of art]'.[284] Viewed through this lens, the Paris Commune (or, indeed, *the right to the city*) can be understood as a struggle of the workers for the fruits of their labour; for control over the city and urban life, their work of art.[285]

Written to commemorate the centenary of Marx's *Capital* in 1967, Lefebvre's seminal essay, 'The Right to the City', is both a *response* to the crisis of the city (particularly his city, Paris) and the impoverishment of the every-day life of the city under capitalism, and a *call* for the right of the working class to remake the city and urban life in a 'less alienated, more meaningful and playful' form.[286] Lefebvre acknowledged this project as utopian, but argued that only 'not very interesting people escape utopianism'.[287] He goes on to note:

> Since the Middle Ages, each epoch of European civilization has had its image of the possible, its dream, its fantasies of hell and paradise. Each period, and perhaps each generation has had its representation of the best of all possible worlds, or of a new life, an important, if not essential part of all ideologies.[288]

The point here is not to take these past visions of utopia as models for the future, but that utopian visions are an important guide for decision-making. As Lefebvre argues, '[w]hy should the imaginary enter only outside the real instead of nurturing reality?'[289]

---

[280] Harvey (n 194) 128.
[281] Ibid 129.
[282] Ibid.
[283] Ibid.
[284] Lefebvre, *Right to the City* (n 194) 4.
[285] Ibid 11; Harvey (n 194) 129–30, 137.
[286] Lefebvre, *Right to the City* (n 194) (particularly 57–64); quote from Harvey (n 194) x.
[287] Lefebvre, *Right to the City* (n 194) 59.
[288] Ibid 66.
[289] Ibid 69.

Lefebvre's essay was written not long before May 1968, when Paris (and the rest of France) once more became the site of insurrection, and first students then workers protested, held general strikes, and occupied universities and factories across the country. Harvey argues this 'is highly significant . . . [since h]is essay depicts a situation in which such an *irruption* was not only possible but almost inevitable'.[290] Like this book, Lefebvre was focused particularly on the geography of revolution, and he emphasised the importance of spaces in the centre of the city where people can spontaneously come together to both reclaim and recreate the city.[291]

At the end of his essay, Lefebvre concludes:

> The right to the city manifests itself as a superior form of rights: right to freedom, to individualization in socialization, to habitat and to inhabit. The right to the *oeuvre*, to participation and appropriation (clearly distinct from the right to property), are implied in the right to the city.[292]

Noting the inherent ambiguity of many of these claims, Harvey argues that 'the right to the city is an empty signifier. Everything depends on who gets to fill it with meaning. . . . The definition of the right is itself an object of struggle, and that struggle has to proceed concomitantly with the struggle to materialize it.'[293] Here the value of prefigurative politics becomes evident. The Paris Commune, Winstanley's colony on St George's Hill, and Kett's camp at Mousehold Heath were all attempts to simultaneously define communal claims to an alternative social order grounded in spatial justice and to materialise such a utopia. To paraphrase Marx, their greatest achievements were their own working existences. These 'liminal social spaces of possibility'[294] created during times of inflection provided a place to perform communal relations to property and to reclaim the power associated with these rights.

Looking forward, Harvey proposes that any political agenda 'must provide answers to three compelling questions'.[295] The first is how to address the problem of 'crushing material impoverishment' in order to give people the freedom to flourish – 'that realm of freedom which begins when the realm of necessity is left behind'.[296] Here we observe that private wealth always results

---

[290] Harvey (n 194) xi.
[291] Lefebvre, *Right to the City* (n 194) 11, 70–1.
[292] Ibid 73.
[293] Harvey (n 194) xv.
[294] Ibid xvii.
[295] Ibid 126–7.
[296] Ibid 127, citing Karl Marx, *Capital Volume III* trans. Ernest Unterma (International Publishers, 1977) 820.

in public poverty, and draw inspiration from the Commune's commitment to *communal luxury*. The second question relates to confronting 'out-of-control environmental degradations and ecological transformations'.[297] In relation to this second question, we argue that confronting environmental degradations requires that we acknowledge the lawful forest is a necessary prerequisite of the right to the city. As we have argued elsewhere, '[n]ot only do forests (and rural commons) have deeper historical roots, but their preservation is essential to the health of the city and its inhabitants.'[298] Finally, Harvey's third question, which is inherently entangled with the first two, involves challenging the 'law of endless capital accumulation' and, with it, the destructive commitment to perpetual growth.[299] With these questions in mind, we shift once more in location to the forests of Australia and a time when city dwellers returned to the forest to reclaim its liberties: communal luxury and ecological sustainability.

---

[297] Harvey (n 194) 127.
[298] Clark and Page (n 188) 29.
[299] Harvey (n 194) 127–8.

5

Ecological Communes

BANGALOW PALM

In the late 1960s, but especially the early 1970s, a significant cohort of young people, many of them tertiary educated,[1] escaped Australian cities in demographically significant numbers. This was the era of the 'back to the land' movement; one of only four occasions in Australia's urbanised colonial history when the unidirectional flow of rural to city migration was reversed.[2] Much scholarly attention has focused on the zeitgeist of this countercultural era, particularly in the United States.[3] In this chapter, we explore how the Age of Aquarius took a distinctly Australian form: the enactment of a rural utopian vision that played out largely on the degraded green hillsides and forested escarpments of the far north-eastern corner of New South Wales.[4] Here, ancient subtropical rainforests had been clear-felled in less than a century, a frenzy of destruction that left a paltry 1

---

[1] Bill Metcalf and Frank Vanclay, *Social Characteristics of Alternative Lifestyle Participants in Australia* (Environmental Information Series, Report No. 1, Griffith University, 1986). The percentage of alternative lifestylers with tertiary or postgraduate qualifications ranged from 32 per cent to 47 per cent, ibid 46. This group is described as having made a 'positive choice' in lifestyle, and therefore 'their commitment remains high', ibid 48. Interestingly, women were more likely to have had university education than men, ibid 49. See also Margaret Munro-Clark, *Communes in Rural Australia: The Movement since 1970* (Hale & Iremonger, 1986) 64, who notes that 32 per cent of rural communards in 1985 had university degrees compared with 4 per cent of the general population.

[2] Lee Stickells, 'Negotiating Off-Grid' (2015) 25(1) *Fabrications: The Journal of the Society of Architectural Historians, Australia and New Zealand* 109 ('Negotiating').

[3] For example, Benjamin Zablocki, *Alienation and Charisma: A Study of Contemporary American Communes* (Free Press, 1980); Timothy Miller, *The 60s Communes: Hippies and Beyond* (Syracuse University Press, 1999).

[4] There were also countercultural settlements established in Western Australia's South West, and in the tablelands of Far North Queensland and behind the Sunshine Coast. But in the hinterlands of the NSW Northern Rivers, the legacy of these times proved enduring, anchored by the National Union of Students' Aquarius Festival held in Nimbin in May 1973, and the subsequent establishment of large numbers of multiple occupancy communities that persist to today, Stickells, 'Negotiating' (n 2).

per cent of remnant cover. This forest, colloquially called the Big Scrub, remains extant at its fringes, confined to the headwaters and ravines of creek valleys and atop the Nightcap Ranges and Jerusalem Creek mountains. But even into the early 1970s, it was a forest that, in its once expansive magnificence, continued to haunt both settler colonial and Indigenous memories.[5]

This chapter begins at this point of inflection, some fifty-plus years ago, when so-called 'longhaired hippies' escaped the enclosures of the city, and in their idealism and youthful naivety, began an experiment in what they saw as a new way of relating to land and the environment. Theirs was a story of forging pioneering communitarian property-holding models, of the establishment of multiple occupancy rural land-sharing communities ('MOs')[6] where new ways of relating to land and an emerging environmental ethos emphasised 'repairing the ravages of previous land use battles and liv[ing] in accord with the natural environment' – not 'battling the bush in fruitless attempts to subdue it'.[7] As noted, this internal migration had its epicentre in the far north-east of NSW, the aptly named Northern Rivers region bordered by the towns of Byron Bay to the east, Lismore to the south and Murwillumbah to the north. This was (and is) a green subtropical land dominated by a microclimate of heavy rains, a myriad of creeks and rivers, rich red soils dating from a long-dormant volcano (named Wollumbin or 'Cloud Catcher' by the local Widjabul people) and verdant rainforests. At the geographic heart of this triangle lies the village of Nimbin, an ex-dairying town re-enlivened in May 1973 as the site of the Australian Union of Students' (AUS) sponsored Aquarius Festival. This was Australia's Woodstock, a ten-day-long celebration of countercultural values and alternative lifestyles that became iconic in the annals of Australian 'hippiedom'. An estimated 10,000 people passed through Nimbin, transforming this quiet country town into a 'trial centre for new ways of living' and forging a 'seminal countercultural episode'[8] in Australia's twentieth-century social history.[9]

---

[5] As a child who grew up in the Northern Rivers in that time, the co-author can attest to these recollections as a matter of personal anecdote.

[6] John Page, 'Common Property and the Age of Aquarius' (2010) 19 *Griffith Law Review* 172 ('Age of Aquarius'); John Page, 'Counterculture, Property, Place, and Time: Nimbin 1973' (2014) 17(6) *M/C Journal*, available at <https://journal.media-culture.org.au/index.php/mcjournal/article/view/900>.

[7] 'Report into the Facilitation of a Rural Intentional Community' (undated), Records of the Aquarian Archive Incorporated Association, Southern Cross University, Lismore.

[8] Stickells, 'Negotiating' (n 2) 110.

[9] Bible says that Nimbin and the 1973 Aquarius Festival, with its countercultural symbols of 'drugs, music and "free love" . . . is still a monument to the spirit of the 1960s', Vanessa

The chapter then shifts to the end of that tumultuous decade, from 1973 to 1979, when many of these same Aquarian communards rallied community support to protect one of the last remaining stands of the Big Scrub. These were the forest protests of Terania Creek, a remote, largely pristine valley across the hills from Nimbin. Activist academic Vanessa Bible describes the forests of Terania Creek in the following terms:

> The Widjabul people of the Bundjalung nation resided in these rainforests. The trees provided food and medicine, and supported recreational, ceremonial and spiritual culture. The land was imbued with lore [sic] and every rock, tree, mountain and waterway had its place within Aboriginal culture. One of these waterways was a creek that flowed down the southern edge of the rim left by the once-enormous shield volcano. Its name was Terania, meaning 'place of frogs.'[10]

Plans by the NSW Forestry Commission to log the Terania rainforest and replace it with plantation eucalypts became known in 1974, and a five-year period of what Bible terms 'polite' protest ensued. Direct action in the face of bulldozers, chainsaws and logging contractors took dramatic yet largely non-violent form in August 1979, and after four weeks of forest protest, logging was halted. Thereafter, judicial and political enquiries into rainforest logging were held, and in 1983, these ancient rainforests were gazetted by the NSW Government to become the Nightcap National Park. Bible claims that the Terania protests of August 1979 constituted 'the world's first successful stand-off between people and machinery in defence of a forest . . . a spontaneous blockade in a style and at a scale never witnessed in the forests'.[11] The end of the 1970s also saw a legitimation of what architectural scholar Lee Stickells terms the region's 'new countercultural geography'.[12] The legalisation of MOs within NSW building and planning codes meant that the realities, norms and values of communal property now had state sanction. Like the Terania protests, this represented a watershed, a shift in relationships with property and land that prefigured *other* ways of doing, and from which the 'old ways' of the pre-Aquarian hills of Nimbin never fully returned.

Bible, *Terania Creek and the Forging of Modern Environmental Activism* (Palgrave Macmillan, 2018) 16.

[10] Ibid 2.

[11] Ibid 3.

[12] Lee Stickells, 'Housing the Farmers of Enlightenment' in I Doucet and J Gosseye (eds) *Activism at Home: Architects Dwelling between Politics, Aesthetics and Resistance* (Jovis, 2021) 235–46, 235 ('Housing the Farmers').

After Terania, the chapter pivots to the island state of Tasmania, where the Franklin Dam protests of the early 1980s garnered national headlines. Here, protests against the proposed building of a hydroelectric project on the Franklin River centred on the battle to save another ancient forest from destruction, this time from flooding, not logging, and with it the high-profile drowning of 2,000-year-old Huon pines. This protest seized the national imagination in ways that past attempts to halt hydroelectric development in Tasmania had not.[13] Colourful scenes of flotillas featured in nightly news bulletins, a protest song became a national hit,[14] and in politics, the heated debate was framed as a contest between constitutional state rights (on the Right) and environmental protection (on the Left). The 1983 federal election swept into power the Hawke Labor government, (one of) whose central policy appeals rested on saving the Franklin, which it did by enacting federal laws to override the state's development plans.

The chapter then fast-forwards thirty years, when several of those same actors from the early 1980s (such as Greens Senator Bob Brown) feature in yet another Tasmanian forest protest. Its setting is to the Franklin's north-east, the eucalypt forests of Lapoinya. The Tasmanian forestry commission, loggers and their proponents challenged the legality of protest on public forestlands, relying on technical interpretations of 'business premises' and 'business access areas' in logging coupes. The dispute became the case of *Brown and Hoyt v Tasmania* (*Brown*),[15] where the High Court of Australia takes us back to the lawful forest and the ancient idea that *at law* forests are not only physical spaces of trees, but also metaphysical spaces replete with a rich fabric of jurisprudence and jurisdiction. Spaces that enable ancient public rights – like a qualified right to protest – to persist into the twenty-first century, freedoms with a provenance that loop back to the Forest Charter of Chapter 2. The High Court decision in *Brown*, perhaps like the under-regarded Terania protests of 1979, or the near-invisible Forest Charter of 1217, underscores how the crucible of protest is fundamental to achieving environmental justice for the forest and a reckoning of spatial justice for its forest citizens.

Finally, this chapter's explicitly Australian context invokes a subject hitherto only indirectly addressed in this book: the legitimacy of colonial settler state sovereignty. Or, at the micro-level, the phenomenon that Carol Rose terms 'ownership anxiety',[16] the 'extremely nervous sentences'

---

[13] For example, Lake Pedder drowned in 1972 by a hydroelectric dam development.

[14] Goanna (Gordon Franklin Wilderness Ensemble), 'Let the Franklin Flow' (April 1983).

[15] [2017] HCA 43.

[16] Carol Rose, 'Canons of Property Talk, or, Blackstone's Anxiety' (1998) 108 *Yale Law Journal* 601 ('Blackstone's Anxiety').

that followed William Blackstone's famous 'sole and despotic dominion' mantra of property, words that cast doubt over 'the foundations for existing distributions' of land titles.[17] In the colonial state, the lawful forest, by its very contested and violent pasts – and presents – is a fraught and complex space. On this colonial forest floor, foundational issues of lawfulness remain unresolved, a modern-day dissonance that impugns claims to 'ownership' founded on historic land thefts and enduring legal fictions of *terra nullius*.[18] In this chapter, our account of these emerging relationships with land – recorded here from the early 1970s onwards and seen through the eyes of communards, protestors and the forest itself – implicates settler relations with Australia's First Nation peoples. This remains an evolving relationship with land that is profound to the spatial justice of Indigenous Australians, and ultimately, as flagged, the legitimacy of the modern Australian nation-state.

The Age of Aquarius, Nimbin 1973

The chapter begins at the dawning of the Age of Aquarius.[19] What was originally a phenomenon of American youth culture exploded across Australian university campuses in the late 1960s, but especially the early 1970s – a point in time when it was arguably on the wane in the United States.[20] As a cultural transplant, the Australian countercultural movement shared many of its American parent's core characteristics: a disillusionment with the 'corporate/ capitalist state' born from the anti-Vietnam War era, a search for individualised meaning that sprang from the 'emptiness' of materialist society, and a nascent environmentalism prefigured by events like Rachel Carson's 1962 publication of *Silent Spring*[21] and the first Earth Day celebrated in countercultural San Francisco in March 1970. As Timothy Miller explains (in the American context):

---

[17] Ibid 602. Rose argues that property scholars may be familiar with Blackstone's famous trope of property, yet few scholars have read beyond it.

[18] See, e.g., Irene Watson, 'Aboriginal Laws and Colonial Foundation' (2017) 26(4) *Griffith Law Review* 469 ('Aboriginal Laws').

[19] The production of the 1967 rock musical *Hair* demonstrates a cross-pollination of American countercultural ideas into Australia – as a representation of the American hippie dream to Australian mass audiences. In 1970, it was seen as 'the most revolutionary event to have occurred in Australia for the preceding three years', Munro-Clark (n 1) 56. Bible describes *Hair* as the 'epitome of 1960s counterculture, condensed into one performance and unleashed upon an Australian audience hungry for such material', Bible (n 9) 13.

[20] Zablocki says that 'the years 1967 and 1968 were the heyday' in the US, Zablocki (n 3) 52.

[21] Rachel Carson, *Silent Spring* (Mariner Books, 1962).

Why did it happen . . . the 'why' is inevitably subjective. Disgust with the direction that American culture had taken – especially with the worship of the almighty dollar – had something to do with it. Psychedelics had something to do with it. The Vietnam war had something to do with it . . . American culture had deeply and uniquely frustrated three basic human desires – for community, for engagement and for dependence . . . finally one can only say that an extraordinary zeitgeist materialized in the late 1960s . . .[22]

In Australia, a key feature of this 'extraordinary zeitgeist' was the urge to escape the city; to follow a collectivist ideal premised on a lifestyle of rural 'voluntary material simplicity',[23] and to live in accord with the precepts of nature. As reflected at the time, '[f]or many people, the way out of a fragmented urban-based life was to abandon its difficulties in exchange for a rural lifestyle.'[24] Or, as more bluntly expressed, 'the city is hardly needed, life should spring out of the country'.[25] As Graeme Dunstan, co-convener of the Aquarius Festival, observed in an open letter to Federal Urban Affairs and Decentralisation Minister Tom Uren in March 1973, the degradation of urban environments is 'a constant reminder that all is not going well with our cities'.[26] Dunstan's manifesto (and invitation for Uren to visit the Aquarius Festival) called for the Commonwealth government to support a 'small scale experiment' in decentralised living whereby a 'community group of 100–200 people' would 'dedicate to building a self sufficient community . . . whose central design features are creative living and ecological survival'.[27] To scholar and 1970s communard Peter Cock, the appeal of communal rural life was the 'sense of simplicity and the freedom to roam' that was central to it.[28]

Statistics of positive rural drift from the period affirmed the zeitgeist. Bill Metcalf and Frank Vanclay cite 58 per cent of 'alternative lifestyle participants' as living in rural areas, 23 per cent in small rural towns, with only 19 per cent based in cities.[29] In the country, 'cheap living costs, afford-

---

[22] Miller (n 3) 67.
[23] Munro-Clark (n 1).
[24] 'Report into the Facilitation of a Rural Intentional Community' (n 7).
[25] Tim Jones and Ian Baker, *A Hard-Won Freedom: Alternative Communities in New Zealand* (Hodder & Stoughton, 1975) 5.
[26] Graeme Dunstan, 'A beginning rather than an END', Open letter to Minister Tom Uren inviting him to visit the Aquarius Festival, *The Nimbin Good Times* 1(1) (27 March 1973).
[27] Ibid.
[28] Peter Cock, *Alternative Australia: Communities of the Future?* (Quartet Books, 1979) 5.
[29] Metcalf and Vanclay (n 1) 66.

able land, warm climates and ... [sometimes] tolerant attitudes'[30] fed the trend. Certainly, after the Aquarius Festival ended in May 1973, and unlike Woodstock where festivalgoers quickly dispersed, 'many Aquarians were so inspired by what they created that they decided to keep the spirit of the festival alive',[31] buying up cheap farms and establishing communes. Stickells notes that the Terania Shire Council (centred on Nimbin) was especially welcoming to 'new settlers', its relaxed building codes encouraging an early wave of rural land-sharing communities.[32] Aquarius Festival co-convenor Johnny Allen similarly (and contemporaneously) observed that Nimbin townsfolk 'talk rather protectively and somewhat warmly of "our hippies"'.[33] For a diversity of reasons, from the staging of the Aquarius Festival to the happy convergence of favourable factors post-festival, the 1970s rural communitarian movement took root in the hills surrounding Nimbin. It signalled an explicit intent to return to a simpler, pre-industrial rural idyll, which in the case of the communards of Aquarian Nimbin, translated into a contextual return to the remnant forests, real and reimagined, of the (then) remote Big Scrub.

The back to the land movement of 1970s Australia also had other key defining features. One was the deliberate decision to embrace communal property models as the medium through which to realise key elements of the zeitgeist –the re-embracing of community values, self-growth and empowerment within the collective, and low-cost aspirations to live closer to nature. One of the earliest (and largest) MOs in the Northern Rivers was Co-ordination Co-operative at Tuntable Falls, located halfway between Nimbin and Terania Creek. It was established in December 1973, and is widely regarded as the first MO set up in the post-Aquarius euphoria. Set on 400 hectares of subtropical rainforest and once run-down farmland, the community comprises a series of 'hamlet clusters' set amongst 'conservation rainforest parcels, individual dwellings, and (variously over time) a

[30] Page, 'Age of Aquarius' (n 6) 175. Similar geographic and social convergences occurred in New Zealand in the Coromandel Peninsula and the Golden Bay region of the South Island, Lucy Sargisson and Lyman Sargent, *Living in Utopia: New Zealand's Intentional Communities* (Aldgate, 2004).

[31] Bible (n 9) 16.

[32] An article in the short-lived *Byron Express* made similar observations in 1973: 'The Terania Shire Council have been pretty good in that there aren't really regulations drawn up for this sort of thing ... So there is a great deal of flexibility on how they want to present it', *The Byron Express* (11 May 1973) 5. Peter Cock likewise observes the Terania Shire was a 'friendly shire council', Cock (n 28) 122.

[33] Johnny Allen, 'Nimbin: Myths, Dreams and Mysteries', in M Smith and D Crossley (eds) *The Way Out: Radical Alternatives in Australia* (Lansdowne Press, 1975).

community hall, store, school, youth club, recycling shed and common lands
set aside for community gardens, woodlots, orchards and dams'.[34] Its initial
legal structure was a co-operative,[35] a consensus-driven statutory model used
by farmer groups that successfully 'internalized common property norms
and practices within a private title system'.[36] This inside/outside dichotomy
is a key conceptual feature of common property models, where the outside
'shell' presents land as being held by an external yet single collective entity,
yet 'inside', this same shell enables and facilitates a sharing of communal
resources in accord with internal rules and common property norms, the
latter described by Rose as virtues of 'moderation, proportionality, prudence,
and responsibility to the others who are entitled to share in the common
resource'.[37] The Aquarian communes, or more accurately, those that matured
and endured beyond their initial honeymoon phase, were particularly successful in adapting private law structures such as the co-operative, private
corporation or settled trust.[38]

Harking back to the common property structures of rural England[39] (of
the type explored in Chapter 4, like the rebel camp at Mousehold Heath,
'the greatest practical utopian project of Tudor England'[40]), these 1970s
property experiments reflect Chapter 1's argument that in times of linear
rupture, *continuities* or distinctive patterns of place recur; and a 'whole world
of comparable instances' opens up (again). Most likely, the land-sharing
communes of the 1970s were a serendipitous coincidence. As Benjamin
Zablocki observes (at least in the United States), there was little appreciation
for precedent: 'There is surprisingly little awareness among present-day com-
munitarians of their historical forebears. Few communes set out deliberately
to emulate admired communal experiments of the past or to avoid repeating

---

[34] Bill Metcalf, *Co-operative Lifestyles in Australia: From Utopian Dreaming to Communal Reality* (UNSW Press, 1995) 80. In 2021, it remains a vibrant community. In 2020, at the height of Australia's east coast bushfires, a volunteer team of CC Protectors saved the community from fiery destruction as the rainforest burned. See also Munro-Clark (n 1) 126–30; Cock (n 28) 121–6.

[35] Co-operatives Act 1923 (NSW).

[36] Page, 'Age of Aquarius' (n 6).

[37] Carol Rose, 'Givenness and Gift: Property and the Quest for an Environmental Ethic' (1994) 24 *Environmental Law* 1, 25.

[38] Page, 'Age of Aquarius' (n 6).

[39] Indeed there was also a harking back to the agrarian imagery of pre-industrial England in journals such as *Earth Garden*.

[40] Jim Holstun, 'Utopia Pre-empted: Kett's Rebellion, Commoning, and the Hysterical Sublime' (2008) 16 *Historical Materialism* 3, 5.

the mistakes of those that failed.'[41] Terry McGee, one of the early founders of Co-ordination Co-operative, put into words the free spirit of the times – and the new settlers' utopian hopes for what was seen as a '*new* way of life':

> Life in our cities is rotten to the core. This is an experiment in a new way of life. The people who join us here have to be prepared to jump off the cliff with the certainty that when they get to the bottom, they will be all right. This is a place for people who like each other and want to help each other.[42]

Finally, a key plank of the early 1970s communitarian movement was the emergence of a global environmental consciousness, which in practice seeded what is argued to be (a distinctly Australian) forest protest movement. A direct line can be drawn from the Aquarius Festival in 1973 to Terania Creek in 1979 and (for the purposes of this chapter) ending with forest protests in Tasmania, a protest culture rich in its lawful significance – and accorded due judicial notice (in its public forest context) by the High Court in *Brown*.

Building on (mostly) American scholarship, like Rachel Carson's paradigm-shifting *Silent Spring*,[43] there was a growing understanding of the planet's finitudes, and a correspondingly palpable sense of ecological foreboding amongst Aquarian communards. Theirs was an early premonition of a post-war society edging ever closer to the precipice of William Gibson's 'Jackpot'. In 1975, it was expressed by a senior Commonwealth government minister as 'a common conviction that something is rotten at the core of conventional human existence . . . an assumption that [humans are] an endangered species, that some combination of forces is threatening [their] capacity to cope'.[44] The response was a communitarian embrace of 'new' ways of relating to the environment, ecologically minded practices that reflected the need to confront an unfolding crisis: the rise of permaculture, recycling, food co-ops, early renewable technologies (many Northern Rivers MOs were 'off grid' in their reliance on primitive solar, wind or water power), or the economies of barter. Publications such as *Grass Roots* or *Earth Garden* extolled the utopian virtues of the 'back to the land' lifestyle, preaching

---

[41] Zablocki (n 3) 43.

[42] McGee, cited in Munro-Clark (n 1) 126 and Cock (n 28) 121.

[43] See also Paul R Ehrlich, *The Population Bomb* (Ballantine Books, 1968); Barry Commoner, *The Closing Circle: Confronting the Environmental Crisis* (Jonathan Cape, 1972). In legal scholarship, environmental law emerged as a distinct field of study, exemplified by a raft of new journals such as the *Ecology Law Journal*, launched at the University of California, Berkeley, in 1975.

[44] Hon. Dr Moss Cass, Australian Minister for the Environment and Conservation, 'Foreword', in Smith and Crossley (eds) (n 33).

self-sufficiency, organic gardening, 'Earth craft' and co-operative agriculture. In words reminiscent of the Diggers and earlier radicals, such journals regularly challenged conventional ways of relating to land:

> Land ultimately cannot be owned by anybody. It is constant while human life is transient upon it. We should leave the land at least as fertile and vigorous as we find it in order not to diminish the chances of future generations.[45]

Co-ordination Co-operative was described as 'the first national symbol of the attempt to create self-subsistence survival communities', a 'self-sufficient ecologically balanced community' with the objective that its members 'live in harmony with nature . . .' and 'free from pollution of air, food, bodies and mind'.[46] The early success of intentional communities such as Co-ordination Co-operative ('by 1975 alone there were a thousand "alternative settlers" in the Tuntable valley'[47]) led to a sudden critical mass of new settlers changing the demographic make-up of the Northern Rivers. The shift was not just political, or in the median age of the population. It also led to the creation of a distinct community of environmentally conscious residents. As Bible opines, the new settlers had 'unknowingly transformed the [region] into a concentrated population of creative hippies, idealists, anti-authoritarians, spiritual practitioners, and environmentalists. The "Rainbow Army" that would stand in defence of the rainforests [in four years] had come together.'[48] At the time, however, the new settlers were not yet aware of the significance of the abrupt demographic change they had wrought, the geographically unique 'convergence of many young people of similar philosophical outlook . . . within a few months'.[49]

This short study of the Australian countercultural movement, with its geographic focus on the far north-eastern corner of NSW, resonates with many themes of this book: the embrace of communal living within common property models, a return to the simple 'virtues' of rurality and living close to the land, an ethos of 'voluntary material simplicity' as participants 'dropped out', and a radicalism based on an escape from conservative social control, collective governance and radical new lifestyles. This was a time of change and disruption, but it was also an era that revisited many 'new' old normals. A zeitgeist had swept Western consciousness during these tumultuous years, a

---

[45] Alternative Canberra, 'Land and the People', *Earth Garden* (July 1977) 34.
[46] Cock (n 28) 121.
[47] Bible (n 9) 17.
[48] Ibid 18–19.
[49] Ibid.

'youth-quake' that for some represented the permanent taking of an alternate fork in the road, and a prescient prefiguring of the ecological ruptures ahead. Omnipresent also was the image and spectre of the forest. It was perhaps no accident that the 1970s rural communards of Australia felt an innate need to live closer to (the fact and the metaphor of) the forest. As Dudley Leggett, founder of the Dharmananda intentional community, reflected in 2010 (speaking of the Terania protests of 1979), 'when the forests were about to be logged, we saw that we'd come there particularly for the forest. To be close to that forest.'[50] Perhaps there is no better way to visualise this collective urge than to describe the site of Co-ordination Co-operative's commune in the hills outside Nimbin:

> It nestled at the foot of 8000 hectares of state forest . . . This forest surrounded the property on three sides. Two-thirds of the property was virgin bush and tropical rainforest; the lower, flatter part was farming and pasture land along a substantial, permanent-running creek, inhabited by platypuses. The upper section of the rainforest extended into the head of the valley to the north. The huge cliff face of the falls stood like a mysterious screen in the deep well of mountain that cradled the valley. It had a secluded nature in an unparalleled setting.[51]

Terania Forest Blockades and Countercultural Geographies

This section, and the next, engages with a core theme of this book: that of protest and its critical histories – set in two ancient Australian forests in the latter half of the twentieth century. Arguably, its two tales are quite different. One remains mostly a hallowed local tale of folklore status. The other, by contrast, is the epochal story of the Franklin, a moment of national consequence that became indelibly etched in Australia's social, legal and political history, and its discourse of lawful protest. Yet their similarities are also plain to see. Each enacted a popular Aquarian exhortation: to 'think global and act local'. Critically, each saved an ancient and endangered forest from destruction.

This section then digresses – chronicling how the late 1970s/early 1980s[52] also represented a watershed for communal property models, and their validation within state planning and building laws. Like many stories in this book, the wider implications of this lawful shift were largely lost to plain

---

[50] Ibid 38.

[51] Cock (n 28) 122.

[52] For example, Prime Minister Bob Hawke speculated in 1984 about setting up a national kibbutz-style scheme similar to the ecological communes of the NSW Northern Rivers, Stickells, 'Housing the Farmers' (n 12) 240.

sight, ignored beyond the 'symbolic and material countercultural terroir'[53] of the Northern Rivers. Yet they should be seen for what they were, the state-sanctioned taking of a different fork in the road, and an ecological aspiration for our propertied relationships with land to better reflect a wish 'to walk more gently on the Earth'.[54]

The Terania protests were unsurprising given their spatio-temporality, transpiring a mere six years after the Aquarius Festival of 1973 and set within earshot of the many ecological communes that were fast transforming the region into an enclave of 'creative hippies . . . and environmentalists'. What scholars such as Bible rail against, however, is how Terania Creek is sidelined, overlooked for its national and global significance as the precursor of protest movements to come. In an unremarked way, the Terania protests are perhaps like Chapter 2's Forest Charter. Or to mix familial metaphors, Terania is the poor cousin of its better-known twin, the Franklin forest protests of 1983. By contrast, the Aquarian links to the Tasmanian forest protests are more attenuated. Again this is unsurprising, given that they occurred a full decade or more after the Age of Aquarius's high-water mark. Yet the line can be seen for what it was (and is), the ongoing consolidation of a green politics into Australia's body politic amidst a growing environmental consciousness that first rose to (modern) prominence as an integral part of the then spirit of the times, the zeitgeist of the Age of Aquarius.

The primary significance of the Terania protests, Bible suggests, is the pioneering tactics and methodologies that it introduced and unwittingly institutionalised into forest protest practice in the years after 1979. For the first time ever, anywhere, protestors at Terania Creek demonstrated new ways around which a community could self-organise and unite to save native forests. And rather than being a weak imitation of international protests, the home-grown Australian forest protest movement was original in its embrace of grass-roots direct action, 'an activist pioneer of global influence'.[55]

The campaign to save the Terania Basin from logging began in 1974, when 'new settlers' uncovered plans by the NSW Forestry Commission to log the 'rain forest' and replace it with eucalypt plantation timber. Loggers from the time heatedly deny this, saying their only interest lay in logging buffer zones of brush-box and hardwoods. However, as articles from 1979 attest, the net effect was the (destructive) same, new logging roads had to be opened up, invasive weeds like lantana would flourish, and the hardwood felling still resulted in consequential damage as logged trees 'slid down the steep slopes

---

[53] Ibid.
[54] Ibid.
[55] Bible (n 9) 1.

to the basin floor some 60 metres away . . . crashing into the rain forest plants and damaging them' as they fell.[56]

The period from 1974 to 1979 saw the newly formed Terania Native Forest Action Group (TNFAG) engage in 'polite' protest, when 'conservationists acted within the boundaries of perceived social norms, using "legitimate" social practices [letter-writing, petitions, political lobbying, raising public awareness] in the hope that respectful conduct would be met with a respectful response'.[57] Yet even during this phase, Bible points to the uniqueness of the TNFAG's decision-making that 'pushed the boundaries and forged new tactics'.[58] Its small membership of seven to ten forest residents drew on their collective 'academic skills' (with backgrounds in environmental science, architecture, engineering, advertising, art and music) to devise new and innovative ways to fight the Forestry Commission.[59] These included challenging the forestry expertise of the Commission itself, and the initiation of 'citizen-science transect surveys' to gain a better understanding of the forest environment under threat.[60] As stated, other responses reflected the creativity of the Aquarian communards who lived in nearby MOs. Musicians composed protest songs for the Terania campaign,[61] and the advertising executive devised a local television advertisement that depicted logging trucks on the same narrow dirt roads as school buses.[62]

By the second half of 1979, political factions within the NSW Labor government had taken entrenched (and opposing) sides in the escalating debate. That this issue had such an outsized influence in faraway Sydney was testament to the TNFAG's lobbying prowess. Timber industry unions, personified by the state's Forestry Minister Lin Gordon, were implacably against the protestors. Other ministers were more sympathetic, including Planning Minister Paul Landa, an instrumental figure in this story for other reasons soon to be canvassed. Landa counselled TNFAG to mobilise wide community support, which they did through alternative society press and at stalls at local markets. Bible claims that TNFAG believed that 'logic would eventually prevail', so when last-minute plans were announced for logging to begin in August, they were unprepared and wrong-footed. What followed

---

[56] 'Terania Creek Protestors Slow Logging', *The (Sydney) Tribune* (29 August 1979) 3.
[57] Bible (n 9) 39.
[58] Ibid 40.
[59] Ibid.
[60] This culminated in the discovery of the rare Nightcap Oak in a proposed logging coupe, ibid.
[61] Paul Joseph, 'Let's Go Down to the Forest' (1976).
[62] Bible (n 9) 46.

was a spontaneous grass-roots response to Landa's invitation, the setting up of a protest camp on the last private property before the state forest began, with equally spontaneous plans for a forest blockade coalescing on the ground as a 'default position'. Within a short time, 300 protestors were *campyng* at the Nicholson property, transforming it into a tent city 'complete with 24-hour running hot water, toilets and a camp oven'.[63] Its *modus operandi* was 'just learning to do on the run'.[64] As is often observed of protest in the lawful forest, the camp was described in festive terms, 'amazingly spontaneous and vibrant' notwithstanding the tensions that lay at its heart. Bible quotes Hugh Nicholson, whose memories of the camp thirty years later recall a place that was 'just spontaneous, . . . people getting together and organising . . . volunteering for the kitchen detail, cooking meals, gathering firewood, providing first aid or child minding . . . I found that really inspiring, that that could happen . . . '.[65]

Of course, Terania's reputed legacy only truly materialised when logging trucks rolled along Terania Creek road on the morning of 14 August 1979. Principles of non-violence, civil disobedience and consensus decision-making collided with loggers and police in large numbers. As the Communist Party's *Sydney Tribune* reported on 29 August, '[t]he NSW Labor government has sent in a special police force [of over 100 officers] costing $250,000 so far. At most, the logging royalties will bring the state $45,000.'[66] The *Australian Women's Weekly* in November published an even lower estimate, reporting that the forest's '1,360 trees [were] worth $25,000 in royalties to the state'.[67] Meanwhile, the Terania Creek Inquiry (headed by retired Supreme Court judge Simon Isaacs) would eventually cost over $1 million.[68]

The blockade at Terania Creek is claimed to be the forerunner of forest protest practices and techniques. Teranians, it was said, 'had unwittingly created history'.[69] After four weeks of stalemate, it also drew a political response, a compromise that stopped logging, initiated the Isaacs judicial enquiry, and eventually saved the Terania Basin with the gazettal of the Nightcap National Park in April 1983. Tree sit-ins, bulldozer obstructions, road sabo-

[63] Ibid 48.
[64] Ibid 49.
[65] Ibid 48.
[66] 'Terania Creek Protestors Slow Logging' (n 56) 3.
[67] Elizabeth Murphy, 'The Battle of Terania Is Not Over', *The Australian Women's Weekly* (November 1979) 11.
[68] The judicial inquiry ran from March 1980 to June 1981.
[69] Bible (n 9) 58.

tages, theatrical protest, so-called 'black wallabying'[70] and other innovative strategies were trialled at Terania Creek – and found successful.[71] This was a spontaneous, organic protest that aligned with the exigencies of the moment, and the evolving spirit of the times. As was observed, '[i]t has the Aquarian touch – light, forward-looking, innovative, portentous, real and relevant to the times.'[72] This story ends in April 1983, when NSW premier Neville Wran announced the state's newest national park at Nightcap. Such timing yields a neat segue for this chapter, from geographic north to south, from the ancient rainforests of Gondwanaland to the ancient Huon pines of north-western Tasmania, and from the ending of one protest to the beginnings of another. But before we leave the region's green, countercultural terroir, there is another parallel (and brief) story to tell: of the ecological communes that were the wellspring of the Terania protests, and how they found their lawful place amidst the state's propertied landscape.

There was an undeniable link between MOs and the Terania forest protests.[73] Many if not most protestors lived in one. It is interesting, therefore, that around the same time, each secured lawful wins. A national park was declared in 1983 four years after the state intervened to stop logging, while a state planning instrument was signed into law in late 1979 that authorised the phenomenon of the rural land-sharing community. Like the inexorable momentum of the Terania protests that saved a forest, intensifying opposition to MOs triggered a corresponding break-through.[74] The amalgamation of Terania Shire into an enlarged Lismore City Council had accelerated a harder line towards MOs. Aerial surveys and increased building inspections led to series of demolition orders being made against unauthorised structures in many of the fledgling MOs. A referendum held in the Lismore local

---

[70] This is the practice of 'flitting through the forest in the intended path of felled trees, hoping that loggers would cease work for safety's sake', ibid 61–2.

[71] Ibid 57. See more generally, 57–80. Bible describes a 'melting-pot of ideas', ibid 103.

[72] Ibid 58, citing Aquarian Festival co-convener Graeme Dunstan.

[73] In a review of the 'bold social revolution' of MOs in 1988, the region's then daily newspaper *The Northern Star* reported that '[t]he issue of logging had also helped inflame the tension and underlying violence towards those living in multiple occupancies. Those trying to stop the logging . . . mainly came from the multiple occupancies which were regarded as havens for hippies, the unemployed, and professional protestors', Kevin Corcoran, 'Learning to Come to Terms with a Different Lifestyle', *The Northern Star* (14 September 1988) 13–14.

[74] The Terania protests played a role in poisoning the well of local tolerance of the new settlers and their communities. Resentment and perceptions of double standards over the enforcement of building codes at its most extreme spilt over into threats of violence: 'There'll be bloodshed if we're not careful . . . a group of angry timber men were all set to go up there with pick handles and clear the hippies out of the forest', Murphy (n 67) 11.

government area in 1979 voted 'no' to future MOs, but in December of that year, Minister Landa intervened. Landa convened a 'hamlet seminar' in Lismore, and shortly thereafter issued a one-off retrospective validation of 'experimental building areas' in certain parts of the Lismore Council area described as inside the 'hippie line'.[75] As Stickells writes, this intervention saw the opening negotiations of a 'new countercultural geography . . . its territories and boundaries encoded into law'.[76] In July 1980, Circular No. 44 was issued by the then Planning and Environment Commission that 'support[ed] multiple occupancies of rural properties in *common ownership* as an appropriate form of settlement in rural areas'.[77]

Stickells charts this shifting legal paradigm through the lens of Bodhi Farm, an MO established by architectural scholar Peter Hamilton and twelve others in 1977. Hamilton's experiment was one of 'commitment to Buddhist practices, voluntary simplicity, [and] consensual decision-making'.[78] Bodhi Farm became embroiled in the escalating 'code wars' between the Lismore Council and MOs, with twelve of its structures deemed illegal and subject to demolition orders. The Bodhi Farmers' journey through this fraught legal landscape ebbed and flowed. Finally, in 1984, the NSW Land and Environment Court upheld their appeal against the council's demolition orders, and the community's lawful status was secured. Bodhi Farm is interesting because of its explicit commitment to communal property. Its 60 hectares of 'largely regrowth forest' is held in trust, with all improvements owned by the community, and individuals reduced to shareholders in the collective. Its (choice of) language makes this clear when applied to its grounded context:

> The buildings used for sleeping, cooking, eating, and so on, were described as 'dwelling segments' and given names such as 'Effort-less' or 'The Glade' to disassociate them from individual ownership. 'Expanded house' was the nomenclature used by the Bodhi Farmers . . . to describe this way of living in close proximity to each other using common facilities. These prosaic domestic adjustments were vital moves in the community's unsteady actualisation of their communal living ambitions.[79]

---

75 Stickells, 'Housing the Farmers' (n 12) 234–5. See also Eddie Buivids, Homebuilder's Association of Nimbin, 'Rural Home Building – A Question of Basic Rights: Learning to Come to Terms' (undated 1988).
76 Stickells, 'Housing the Farmers' (n 12) 235.
77 Page, 'Age of Aquarius' (n 6) 188 (emphasis added).
78 Stickells, 'Housing the Farmers' (n 12) 233.
79 Ibid.

The countercultural geography of the NSW Northern Rivers is a mix of voluntary ecological simplicity, utopian communitarianism, and a feisty readiness to defend its forests. In engaging with a critical history of property, protest and spatial justice, its tale resonates as one that goes against the linear grain, a fertile space and ever-dynamic place of the lawful forest.

The Lawful Forests of Tasmania: Of Peaceful Protest and High Court Vindication

The lawful forests of Tasmania have a similarly rich history of protest. As the High Court majority observes in *Brown*, '[t]here is . . . a particular historical, social and legislative background to forest operations and public access to forests in Tasmania.'[80] And as Justice Gageler consequentially notes, a 'long history of on-site political protests on Crown land in Tasmania, directed to bringing about legislative or regulatory change on environmental issues, beginning with the protest activity between 1981 and 1983'.[81]

This 'background', and its 'long history of political protest', has its modern beginnings in the failed attempt to save Lake Pedder in 1972, renowned for its 'ancient still waters and sparkling beach of pink quartzite', lost to a controversial dam built by the Tasmanian Hydro-Electric Corporation (HEC).[82] The fate of Lake Pedder led the Whitlam Commonwealth government to establish the National Estate, 'for the first time recognising the natural environment as an integral part of Australia's heritage'.[83] Likely, it also 'galvanised a movement and made victories elsewhere possible. And it was the prompt that led to the formation of the Australian Greens.'[84]

When plans were announced by the HEC for the damming of the Franklin below the Gordon River in south-west Tasmania, there was a fierce determination not to repeat the mistakes of Lake Pedder. As Clive Hamilton drily notes, 'the planning for the Franklin blockade in the summer of 1982–1983 was long and thorough'.[85] And so, the clock started ticking on Justice Gageler's timeline, beginning a decades-long continuum of Tasmanian forest protest. Interestingly, in a nod to the Terania protests, Hamilton cites the tactic of establishing 'affinity groups of 6–10 protestors for mutual support', and the use of music and song to 'express feelings, lift spirits and build resolve'[86] as positive lessons learnt from the Terania experience.

[80] *Brown* (2017) 261 CLR 328, 341.
[81] Ibid 346, 387.
[82] Clive Hamilton, *What Do We Want? The Story of Protest in Australia* (NLA Press, 2016) 168.
[83] Bible (n 9) 23.
[84] Hamilton (n 82) 168.
[85] Ibid 176.
[86] Ibid.

Integral to the Franklin protests was the leadership of the Tasmanian Wilderness Society and its charismatic director, Dr Bob Brown (later Senator and Leader of the Australian Greens). Brown inspired 6,000 protestors to flock to the small town of Strahan, where a protest camp was set up downstream from the dam site. Over that summer, protestors were dispersed from Strahan to a blockade site named 'Greenie's Acres' (set up on a sympathetic landowner's private farm), and from where flotillas of inflatable 'ducks' set sail daily to block the passage of water-borne earthmoving equipment, and otherwise disrupt the dam's construction.

Organisers soon realised that to succeed the protest had to gain and hold national attention. Both major Tasmanian political parties and the state bureaucracy itself were classic examples of 'agency capture', whereby the powerful HEC was in effect the *de facto* state government. Winning the hearts and minds of Tasmanians was not enough. An onsite communications centre, set up by the Wilderness Society, broadcast the protests to national media outlets, and nightly news bulletins across that summer featured the colour, drama and aesthetics of protest. The song 'Let the Franklin Flow' filled the airwaves of metropolitan FM radio stations and popularised the cause. Within months, public sentiment firmed strongly in favour of 'No Dam'. That sentiment was concretised when the conservative prime minister, Malcolm Fraser, called an early federal election for 5 March 1983, and on the back of a promise to save the Franklin, the Hawke Labor government won decisively. The new government moved quickly to honour its election commitment, passing legislation to stop the dam's construction, which the High Court in the *Tasmanian Dams Case*[87] subsequently confirmed as within constitutional power.

The Franklin protests represented a watershed in the Australian environmental movement. As Hamilton posits, 'the Franklin blockade earned the most conspicuous place in the folklore of Australian environmentalism'.[88] Its political and legal vindication, and the integral role non-violent protest played in its success, proved a turning point. It encouraged other forest protestors to adopt similar tactics and objectives, from forests elsewhere in Tasmania to long-running battles over the forests of the NSW south-east and Victoria's Gippsland, which raged throughout the remainder of the 1980s and into the 1990s.

The lesser-known protests at Lapoinya in 2016 form part of this judicially recognised continuum of forest protest dating from the Franklin. At Lapoinya,

---

[87] *Commonwealth v Tasmania* (1983) 158 CLR 1.
[88] Hamilton (n 82) 176.

(now retired Senator) Brown along with fellow protestor Jessica Hoyt were arrested for entering a designated 'business access area', which was, in substance, an off-limits logging coupe proclaimed as such under the Tasmanian Workplaces (Protection from Protestors) Act 2014 ('the Protestors Act'). The Protestors Act, as its title implies, sought to 'protect' forestry workers from protestors in Tasmania's public forestlands, creating zones in which protestors were prohibited to enter and remain under the guise of 'workplace safety'. Dr Brown and Ms Hoyt were arrested at different times when they entered 'the Lapoinya Forest for the purpose of raising public and political awareness about the logging of the forest and voicing protest to it'.[89] Challenging the state Act's constitutionality, the plaintiffs argued that the implied freedom of political communication in the Australian Constitution had been 'impermissibly burdened' by the Protestors Act. Given the High Court's emphasis on context, the following is Brown's description of Lapoinya, written for a blog on the Patagonia website:

> Lapoinya is a huddle of farms in the northwest of Tasmania, Australia's island state. Its rolling hills have a patchwork of lush pastures, ploughed fields and copses of trees. At the heart of the district was the Lapoinya Forest, a couple of hundred acres of wildlife-filled rainforest, eucalypts and ferneries with the crystal-clear Maynes Creek – a key nursery for the world's largest freshwater crayfish – running through it. The forest was prime habitat for other threatened and rare creatures like the Tasmanian devil and giant wedge-tailed eagle.[90]

The High Court agreed with the plaintiffs, holding that the Protestor's Act significantly burdened the implied freedom of political communication.[91] The state Act was therefore not compatible with the maintenance of a constitutionally prescribed system of government, and declared invalid.[92] But what *Brown* represents for the lawful forest goes beyond the case's apparent significance for the implications of Australian constitutional law. With a nod to the US public forum doctrine, *Brown* acknowledged an ancient *place-based*

---

[89] *Brown* 340.

[90] Bob Brown, 'Australian High Court Upholds Peaceful Protest', *Patagonia Roaring Journals* (24 April 2018), available at <https://www.patagonia.com.au/blogs/roaring-journals/australian-high-court-upholds-peaceful-protest?constraint=tarkine>. Patagonia was the largest donor to the fund started to cover legal costs.

[91] The Court compared the unencumbered rights of school children on excursion or hikers to enter the forest with that of protestors, which as a group had been singled out, *Brown* 394–5.

[92] Cristy Clark and John Page, 'Of Protest, the Commons and Customary Public Rights: An Ancient Tale of the Lawful Forest' (2019) 42(1) *UNSW Law Journal* 26, 30–5.

liberty, a customary right to enter public forests (dependent on the public-ness of the forum) expressed by the majority judgment as 'an expectation on the part of the public in Tasmania, residents and visitors alike, that they may access forest areas and that that expectation should, so far as reasonably practicable, be met'.[93]

What is equally significant is the antecedence of this right. The Court, but for one judge,[94] was prepared to leave its origins in the ancient forest, stat-ing that later state Forestry Acts 'recognised' a long-standing public expecta-tion to access public forests and protest. Ironically, it was an amendment to the Forest Management Act in 1983 (enacted post-Franklin) that especially piqued the Court's interest. The 1983 amendment inserted a '"situation of trespass" that was otherwise absent . . . [a change which] revealed a proper-tied truth about the public right, much like thunder informs us of unseen lightning'.[95] This amendment validated the state's authority to remove pro-testors obstructing logging operations, a police power to enforce a trespass right that previously did not exist. This 'recognition approach' implied '[the right's] general or common law origins – a provenance that predated state forestry statutes whose scheme was and is regulatory, not [one] creative of rights'.[96] Or as Brown simply put it, an entitlement which 'guaranteed the public a *time-honored* access to the forests'.[97] Finally, and not understating its significance, this public customary right not only sanctioned access, or the kind of agrarian rights that the Forest Charter of 1217 envisaged for its 'good and lawful men', but also a right to protest, updated in the early twenty-first century to include 'peaceful protest activities such as . . . filming and investigation'.[98]

As noted elsewhere, 'discussion of the public forest right was subliminal to the higher order constitutional issues. One was ratio, the other obiter; one determinative, the other its context.'[99] Yet the High Court's reference to this quaint notion of a 'public forest estate' infuses *Brown* with a long jurisprudential history, one found on the lawful forest floors of Lapoinya, as much as it once lay in the Norman forests of Chapter 2.

---

[93] *Brown* 345. Qualifications on the right were matters of common sense, to comply with the reasonable safety and operational requirements of foresters.

[94] Justice Edelman deemed it a 'statutory licence' under the Forest Management Act 1983 (Tas.), *Brown* 490.

[95] Clark and Page (n 92) 32.

[96] Ibid 33.

[97] Bob Brown (n 90) (emphasis added).

[98] *Brown* 490, per Edelman J.

[99] Clark and Page (n 92) 34.

Contested Sovereignties, Ownership Anxiety and the Settler Colonial State

In 1971, the traditional owners of Country in the Gove Peninsula of Australia's Northern Territory, the Yolngu people, brought an action in the Northern Territory Supreme Court, seeking compensation for an unauthorised acquisition of property under section 51(xxxi) of the Commonwealth Constitution. The property, it was claimed, had been wrongfully acquired, bauxite-mining leases granted by the Commonwealth to the miner Nabalco without the traditional owner's consent or payment of compensation on just terms. The case, *Milirrpum v Nabalco Pty Ltd and the Commonwealth of Australia* (*Milirrpum*),[100] is often referred to as Australia's first land rights case. The plaintiffs had the misfortune that their litigation ended up before a single judge of a territory Court, a white male bound by the precedent of higher courts that deemed Australia *terra nullius* (empty and uninhabited), the legal fiction that grounds the Crown's 1788 claim to sovereignty of the Australian continent. Equally unfortunate, the Yolngu people's case was litigated before its time, some twenty-one years before the High Court admitted to the recognition of native title interests within the Australian common law in *Mabo v Queensland (No 2)*.[101] It was also premature for reasons stemming from the law's longer-running lack of imagination, a structural myopia that fails to see property beyond its private, individualised modality.[102] Legal conceptions of 'property' in early 1970s Australia struggled to accommodate pluralist understandings – like Aquarian communal ownership. The law certainly did not envisage interests in land sourced in the laws and customs of its Indigenous peoples. A blinkered adherence to the common law doctrine of tenure demanded that all interests in land be sourced from a Crown Grant.[103]

---

[100] (1971) 17 FLR 141.

[101] *Mabo (No 2)* (1992) 175 CLR 1 recognised for the first time native title rights and interests within the Australian common law. Native title was seen as a 'burden' or 'encumbrance' on the Crown's so-called 'radical title', leading to its vulnerability to extinguishment. The High Court observed that any other formulation was practically impossible, it being 'too late in the day to fracture a skeletal principle of Australian land law', namely the doctrine of tenure dating from eleventh-century England.

[102] Blackburn J briefly hinted at the common law's ancient plurality, citing 'customary rights, manorial rights, rights of fishery' as 'ancient rights provable by so-called reputation evidence'. However, in the case of the Yolngu's ancient laws, 'the common law had no knowledge of, and could not recognize, rights of the kind which the plaintiffs are seeking to enforce', *Milirrpum* 155.

[103] The Yirrkala bark petitions presented to the Court as lawful evidence of ownership were

This was the legal milieu that reflected the social and political landscape of the 1970s.[104] For the purposes of this chapter, this was likewise the state of the Anglo-Australian common law at the dawning of the Age of Aquarius. Today, this legal space, and especially the intersection of colonial and First Nations law, is far richer, pluralist and contested. Indigenous legal scholars such as Irene Watson, Mary Graham and Aileen Moreton-Robinson, amongst others,[105] argue powerfully that Indigenous relationships with land remain defined (and constrained) by colonial law, a 'weak . . . legal system that has no concept of land and the connections we [First Nations People] have to each other'.[106] As Graham explains, the two worldviews are profoundly different. For Indigenous Australians, 'the land is the law, a sacred entity, not property or real estate, the great mother of all humanity'.[107] Watson is especially critical of native title, calling it out as a construct of the settler state, 'a bullshit construction of Aboriginal recognition [and] a gravy train for the recognition of rights to land'.[108] Native title corrals ancient laws to 'assimilate' with the dominant legal hegemony, which in turn mischaracterises it as a 'burden' or 'encumbrance' on the Crown's superior radical title, thereby leaving it vulnerable to extinguishment. The solution, Watson argues, is profound and fundamental:

> It is time for the state and other historic and ongoing agencies to give back the lands you have stolen from First Nations Peoples. It is time to fully reject terra nullius, to de-centre colonial power and to centre Aboriginal law's authority over place. At the centre of this is the call to restore our laws that have been breached by the colonial project and to begin again.[109]

---

not admitted into evidence. Today they are celebrated as cultural and legal treasures of the nation.

[104] For a wider narrative, see Ken Mackie, Elise Bennett Histed and John Page, *Australian Land Law in Context* (Oxford University Press, 2012) 62–132.

[105] A useful summation can be found in Greta Bird and Jo Bird, 'First Nations Cultural Loss: Whiteness and the *Timber Creek* Judgment' (2021) 1(1) *Legalities: The Australian and New Zealand Journal of Law and Society* 68.

[106] Watson, 'Aboriginal Laws' (n 18) 471.

[107] Mary Graham, 'Some Thoughts about the Philosophical Underpinnings of Worldviews' (1999) 45 *Australian Humanities Review* 181.

[108] Watson, 'Aboriginal Laws' (n 18) 471.

[109] Ibid 469. Elsewhere, Watson writes of *terra nullius*, '[t]here is no death of terra nullius . . . The celebration of the death of terra nullius is a farce: a collective act of schizophrenia, a false-hood, a conspiratorial lie, which has lulled the Australian psyche into a fantasy myth that here had been in the Native Title decision a recognition of Indigenous peoples' rights', Irene Watson, 'Indigenous Peoples' Law-Ways: Survival against the Colonial State' (1997) 8 *Australian Feminist Law Journal* 39, 48.

This is the macro-picture of contested sovereignty in the settler state, the colonial project that 'trembles at the Empire's edge'[110] and confronts a growing acknowledgement (in the legal academy and beyond) that this land 'always was, always is, and always will be Aboriginal land'.[111] It also depicts a micro-picture of what Rose terms 'ownership anxiety', the doubts that landowners harbour as to the provenance of their title. This is the lesser-known postscript to Blackstone's oft-cited mantra of property as 'that sole and despotic dominion which one man claims and exercises over the things of the external world in total exclusion of the right of any other . . .'. It is the curious follow-up, '[p]leased as we are with the possession, we seem afraid to look back to the means by which it was acquired, as if fearful of some defect of title.'[112]

As Chapter 1 argues, citing geographer Tim Cresswell, the stories of property are in various turns dominant and subordinate. This seems to be the two-way bet that Blackstone wagers. The dominant legal narrative of the early 1970s, affirmed by *Milirrpum*, was, however, not the *only* story told at the time. In other reports of case law, there was the faintest hint that the High Court was waiting for the 'right' native title case to present itself. Beyond the law, the later scholarly work of Watson found a rawer edge in the then writings of Aboriginal activists.[113] Even in *Milirrpum*, the precedent-constrained Justice Blackburn conceded in *obiter* that if ever the claimants had 'a subtle and elaborate system . . . a system of law no less, such that if ever a system could be called a government of laws and not of men, it was theirs'.[114] However, as a viable narrative, these subordinate alt-claims to 'property' lacked apparent doctrinal gravitas. Where was the right to exclude, or the right to alienate in the Yolngu's relationship with their land?[115] At the time, the 'gap was [simply] too great'.[116]

In the decades since, subordinate narratives have gained traction. A sophisticated native title jurisprudence has evolved alongside the common law, and enriched property with its pluralism. And as scholarship such as Watson's shows, subordinate stories have also worked to subvert the dominant

---

[110] Ronald Dworkin, cited in Bird and Bird (n 105) 83.
[111] For the history of this phrase see, e.g., <https://australian.museum/learn/first-nations/alwa ys-will-be-aboriginal-land/>.
[112] Blackstone, cited in Rose, 'Blackstone's Anxiety' (n 16).
[113] For example, Jimmy Widders, 'Black Alternatives: Aborigines in the Seventies and Beyond', in Smith and Crossley (eds) (n 33).
[114] *Milirrpum* 267.
[115] Ibid 272–327.
[116] Mackie et al. (n 104) 71.

paradigm.[117] In May 2017, the Uluru Statement from the Heart issued by Australia's First Nations Peoples called for 'Voice, Treaty, Truth' and the sovereignty of First Nations, and the recognition of sovereignty that has never been ceded. As the Statement reads, '[o]ur Aboriginal and Torres Strait Islander Tribes were the first sovereign Nations of the Australian continent and its adjacent islands, and possessed it under our own laws and customs.'[118] Some state governments have begun tentative processes of 'treaty talk' or *Makaratta*,[119] but as yet there has been no shift at the national level.

It is beyond this book's scope to engage broadly with this vast topic. Instead, this chapter takes a more circumscribed approach, viewing the landed relationships between coloniser and indigene through two thematic lenses relevant to the chapter. The first explores how 1970s communards and forest protestors began to critically view the colonial project, drawing on contemporaneous accounts that record their shared encounters. The second gives the forest agency, a comparative exercise from analogous settler states (New Zealand and Canada) that illustrates how the metaphysical forest is conducive to breathing life into multiple legalities and co-sovereignties. Or, as Estair Van Wagner describes, 'the possibility of a legally pluralistic forest where multiple regimes of human–forest relations [may] co-exist'.[120]

The first stirrings of a paradigm shift in race relations in Australia came with the Freedom Ride of 1965, a two-week bus journey through rural NSW by students from the University of Sydney. Inspired by civil rights activists in the US South, the Australian freedom riders sought to emulate their name-sake's success by highlighting racist practices and the ongoing dispossession of Indigenous Australians living on the fringes of white settlement. Hamilton says the freedom riders' strategy was to provoke 'creative tension', challenging long-held prejudices and providing Indigenous disadvantage with a national platform.[121] The freedom riders emerged from a radicalising gene pool on Australian university campuses; the same activist foment that gave life to the Australian Union of Students' led Aquarius Festival eight years later. This was

---

[117] Sarah Keenan, *Subversive Property: Law and the Production of Spaces of Belonging* (Routledge, 2015).

[118] The Uluru Statement from the Heart, available at <https://ulurustatement.org/the-state ment>.

[119] For example, the Noongar (Koorah, Nitja, Boordahwan) Recognition Act (WA) 2016 has been described as the first treaty-like step by any government in Australia's history, Bird and Bird (n 105).

[120] Estair Van Wagner, 'The Legal Relations of "Private" Forests: Making and Unmaking Private Forest Lands on Vancouver Island' (2021) *The Journal of Legal Pluralism and Unofficial Law* 10, DOI: 10.1080/07329113.2021.1882803.

[121] Hamilton (n 82) 100.

the little traversed, little understood context in which Aquarian–Indigenous relations first arose.

In Nimbin, in early 1973, Aquarius Festival organisers came to learn that the site of the Festival was sacred land, a men's initiation site for the Bundjalung people, while the Nimbin Rocks, a haunting presence to the town's south, was a burial site of 'clever men' or 'we-angali'. Women, it was said, 'could not survive there'.[122] In what Bible describes as equal parts naivety and cultural ignorance, the organisers 'thought this was a "curse" that could be removed, so they set out to find a witch doctor – revealing just how little white Australians knew of Aboriginal spirituality in 1973'.[123] However, subsequent dealings with Bundjalung elders slowly changed perspectives and shared understandings. At the Festival, a Welcome to Country preceded the opening, claimed to be the 'first ever' Welcome to Country for white people, while federal funding helped bring 800 Aboriginal people to the festival, including visitors from Pitjantjara in South Australia and Yirrkala from the Gove Peninsula, home of the *Milirrpum* litigation. With the benefit of nearly forty years' hindsight, key Aquarians describe Indigenous participation as 'a truly remarkable part of the festival'. Local Indigenous leaders likewise appreciated the Aquarian zeitgeist, Ruby Langford Ginibi opining that new settlers were drawn to the region for 'the magic of the land . . . an Aboriginal spirituality' that matched the spiritual quest of 'all the hippie-type people'.[124]

However, countercultural efforts to challenge the colonial legacy were not a perfect record of linear progress. Leading countercultural monographs from the 1970s and shortly after bear little or no contemporaneous record of Indigenous impact on the countercultural movement, or vice versa. Margaret Munro-Clark's *Communes in Rural Australia: The Movement since 1970* (1986) makes no reference to Aboriginal Australia, while Peter Cock's *Alternative Australia* (1979), in its sole reference, makes the extraordinary (and cringey) claim that police raids on communes in search of illegal drugs were 'reminiscent of officialdom's attacks on Aborigines a century before': 'Counter-culture communities, particularly in northern Australia, were more severely suppressed than Aborigines in recent times. Aborigines were permitted with government subsidies to live in humpies; hippies were not.'[125]

Cock conversely acknowledges that '[a]lternative seekers have in their lifestyle and their values more in common with Australian blacks than either

[122] Bible (n 9) 24.
[123] Ibid.
[124] Ibid 26.
[125] Cock (n 28) 244.

do with the majority of whites and their culture.'[126] Contrast these mixed messages and poor analogies with New Zealand, where an official land policy, the ohu scheme, was established in 1974 to encourage groups of New Zealand citizens to set up alternative communities in rural areas based on co-operative common ownership models. Prime Minister Norman Kirk recognised the scheme's value in spiritual and social terms, 'to reconnect people with the land and to give them a chance to develop alternative social models'. Kirk's Lands Minister Matiu Rata especially saw its 'strong Maori spiritual dimension' and the capacity to reconnect urban Māori with their *whaungatanga* (or kinship). 'Ohu' replaced 'kibbutz' in the language of early working parties; the word in Māori means 'joining together to do the work'.[127] The scheme involved the granting of long-term leases on unalienated Crown land, with rental calculated at 4.5 per cent of unimproved capital value. While the scheme ultimately had 'a Maori Minister in charge, [and] a Maori name', by 1975, it had 'as yet few Maori participants'.[128] Its passing success, however, should not diminish its utopian aspirations, articulated by Minister Rata as a genuinely New Zealand experiment with the highest motives:

> The over emphasis on the gross national product, perpetual greed, speculation, profiteering, unethical practices and the cult of individualism can only result in the further alienation of those who seek a return to community and group feelings. I share with other Government ministers the hope that ohu communities in some way may . . . recapture anew the deep links of people and land.[129]

By 1979, and back in Australia at the Terania forests, Aquarian communards were reflecting on their maturing relationship with local Indigenous communities. There was a sense that 'we were proving to them that we truly cared for [C]ountry, and that they could link with us'.[130] Bible says that 'Terania was of tremendous importance for the coming together of Aboriginal and white Australia; the campaign [being] "unique" for the unusual collaboration of [I]ndigenous and non-indigenous in taking on authority.'[131] It set a precedent for involving local elders in future forest protests, providing critical Indigenous knowledge such as the whereabouts of ancient artefacts and sacred totems. Activist Burnum Burnum addressed 2,000 people at a

---

[126] Ibid.
[127] Jones and Baker (n 25) 121.
[128] Ibid 133.
[129] Ibid 131.
[130] Bible (n 9) 24.
[131] Ibid.

TNFAG public awareness day at the protest site in August 1979, where he referred to 'Teranians as the "new aborigines", recognising the connection they felt with the land'.[132]

Such is the patchy record from the Aquarian era. This chapter's alternative perspective is to give the forest agency, and to reflect how – as especially metaphysical spaces – forests may be productive of multiple legalities, Indigenous and colonial. This is an interrogation that departs from the coldly rational; where forests are endowed with enchanted, supernatural-like qualities, and timespans have little regard for human brevity. In a sense, this too was a (delayed) spin-off of the Aquarian zeitgeist, the kind of analogous 'hippie-type' spirituality that Ruby Langford Ginibi earlier referred to. In Aotearoa New Zealand, the Te Urewera forests in the central North Island were de-gazetted after sixty years as a national park, and re-enlivened as a separate legal entity under the Te Urewera Act 2014 'with all the rights, powers, duties and liabilities of a legal person' (section 11). In effect, the fee simple title of the land was vested in the forest, such that owner and land became merged into one. The Act defines Te Urewera using the language of the *other*, as 'ancient and enduring, a fortress of nature, alive with history; its scenery . . . abundant with mystery, adventure and remote beauty' (section 3(1)). With its imaginative drafting, the Act sees Te Urewera as at once physical and metaphysical. Its physicality may be 'treasured by all for the distinctive natural values of its vast and primeval forests', its 'indigenous ecological systems', 'historical and cultural heritage', and as 'a place for outdoor recreation and spiritual reflection' (section 3(8)), yet its statutory reach goes beyond the material. For the local iwi, the Tūhoe, Te Urewera is the 'heart of the great fish of Maui, its name derived from Murakareke, the son of the ancestor Tūhoe', their 'place of origin and return, their homeland' and an expression of 'Tūhoe culture, language, customs, and identity' (section 3(4), (5) and (6)). Te Urewera is a distinctly New Zealand accommodation of the nation's bi-cultural identities, a selective rejection of the universality of settler constructs (like national parks), and the incorporation into statute of an Indigenous worldview of natural places as 'ancestral persons'.[133]

Such legal innovations represent a decentring of colonial law, and a concomitant 're-centering of Indigenous legal orders',[134] one that clears

---

[132] Ibid 25.

[133] Martuwarra RiverOfLife, Alessandro Pelizzon, Anne Poelina, Afshin Akhtar-Khavari, Cristy Clark, Sarah Laborde, Elizabeth Macpherson, Katie O'Bryan, Erin O'Donnell and John Page, 'Yoongoorrookoo: The Emergence of Ancestral Personhood' (2021) *Griffith Law Review* DOI: 10.1080/10383441.2021.1996882.

[134] Jones and Baker (n 25) 47.

theoretical space for multiple jurisprudences and (the possibilities of) coexisting sovereignties. Although the trend of conferring legal standing on things of the natural world[135] may be groundbreaking, it is not without critique. For Māori scholar Carwyn Jones, it still 'confirms [in New Zealand at least] that Māori legal traditions will not be recognized on their own terms, but only through the closest equivalent . . . deploying a mechanism from the common law that is deemed by the Crown to be a close enough match'.[136]

Similar considerations arise in Canada. In two separate essays, Kirsten Anker and Estair Van Wagner explore the connections between the metaphysical forest and its sylvan, otherworldly capacity to unsettle hegemonic assumptions and shift lawful paradigms. In the forests of urban Montreal, Anker contemplates 'what it is to take forests as law . . . and the law as forest'.[137] Rejecting 'false' binaries of human/non-human, culture/nature, mind/matter, Anker adopts the 'syntax of . . . metaphor [and] the realm of imagination. Law as forest. Forest as law.'[138] Anker observes how 'Indigenous forms of deep participation in ecological process suggest that the mythos of storied places is more apt to account for grounded jurisprudence than logos.'[139] She concludes that they 'express a truth of a participatory consciousness, in which spirits are a phenomenon produced by the interaction of human minds with other self-organising properties of the world'.[140] The empirics of place and law are a coalescing of rational deduction and metaphysical instinct, divined by 'stories, ceremony, visions, dreams, and walking the land in a mindful way'.[141]

---

[135] Cristy Clark, Nia Emmanouil, John Page and Alessandro Pelizzon, 'Can You Hear the Rivers Sing? Legal Personhood, Ontology, and the Nitty-Gritty of Governance' (2019) 45 *Ecology Law Quarterly* 787.

[136] Carwyn Jones, *New Treaty, New Tradition: Reconciling New Zealand and Māori Law* (Victoria University Press, 2016), 60, 98. Similarly in Canada, Kirsten Anker welcomes the advent of rights of nature discourse, but like Jones, finds it lacking. It fails to 'destabilize' the 'disenchantments' of law, the dualisms of culture and nature, mind and matter, and so on. Entitles such as forests are not genuine actors in themselves, but rather are anthropomorphised by investing them with legal personhood, Kirsten Anker, 'Law As . . . Forest: Eco-logic, Stories and Spirits in Indigenous Jurisprudence' (2017) 21 *Law, Text, Culture* 191, 194, 207.

[137] Anker (n 136).

[138] Ibid 192.

[139] Ibid 194. Anker says the common law is much the same, 'lessons drawn from human drama', ibid 199.

[140] Ibid.

[141] Ibid 207.

To the west, on Vancouver Island, vast private forests ostensibly confer on their corporate owner property rights of exclusion and agenda setting.[142] Yet these forests are also the subject of Indigenous legal orders, worldviews where 'a direct enduring relationship with the land' is pre-eminent.[143] Van Wagner says that 'examining Indigenous forest relations . . . requires looking beyond the human to explore the agreements and the relationships between humans and the *more-than-human world* of the forest, including the metaphysical realm'.[144] There, property relations have spiritual and ancestral dimensions, where certain places are recognised under Indigenous law as First Ancestors and serve as 'living legal scholars'.[145] The lawful consequence of incorporating a 'more-than-human' perspective into private freehold is to unsettle Anglo-Canadian law and its notions of exclusivity. Indeed, Van Wagner argues, it questions whether forests can ever be exclusively 'private'.

In so unsettling, the metaphysical forest implicates the possibilities of legal pluralism, and casts a bright light on the unresolved 'ownership anxiety' that taints title in the colonial state. Like Ginibi's 'magic of the land', or the collective urge of 1970s Aquarians to live near the forest, the 'more than human' dimensions of forests have long spoken to us in a timeless, otherworldly language.

Conclusion

Telling tales of the peripheral, of storied accounts lost to plain sight, forms a core rationale of this book. After all, its chief metaphor, the lawful forest, routinely suffers the same fate. Such under-reportage belies the significance and, importantly, the instructive legacies of the lived experiments of the *other*. In their recurring convergences – the random *throwntogetherness* of *heres-and-nows* – places like Terania Creek, Co-ordination Co-operative or Lapoinya are best seen as moments of societal inflection, signposts towards alternate forks in the road. They point to *other* ways forward, directions that, to our chagrin, we have too little followed. By contrast, well-known tales like the Franklin protests paint a clearer picture of a disrupted continuum, an erstwhile linear progression of improvement and enclosure that goes awry,

---

[142] Larissa Katz, 'Exclusion and Exclusivity in Property Law' (2008) 58(3) *University of Toronto Law Journal* 275.

[143] Van Wagner (n 120).

[144] Ibid 2 (emphasis added).

[145] Ibid 4. 'The need to maintain respectful relationships takes on particular significance in the context of the loss of forestlands, which are relied upon to fulfill a range of social, spiritual and economic needs. Personhood is extended beyond humans to include both "natural" beings such as plants, animals, rocks and supernatural beings', ibid.

'progress' knocked off its trajectory. The Franklin protests engineered a shift in direction in Australian forest preservation, a tack that set us on a different path.

The Age of Aquarius left a lasting imprint fifty-plus years on, in the ecological communes that dot the green Northern Rivers landscape, its utopian literature, protest-era music and bespoke communitarian architecture.[146] It tentatively opened youthful eyes to the brutal truths of settler colonial history and unacknowledged Indigenous dispossession. Aquarians likewise engaged with the forest, protecting its physicality, while intuiting its metaphysicality. Such forest encounters place the 1970s in perspective, as a passing telescoping of time in an ancient, timeless land – a land that Australia's Indigenous peoples have called home for millennia. Finally, the era engendered an environmental consciousness that continues to grow and permeate the mainstream. In its wake, a vibrant protest culture has taken root and been accorded judicial notice, a culture that saved Terania Creek and the Franklin River from the trajectories of enclosure and improvement, and their forests' senseless destruction.[147]

In closing, this chapter underscores how an ethos of voluntary material simplicity became the leitmotif for the times. Aquarians dropped out, forest protestors defended simple ancient liberties, and kinder ways of relating to land and the environment took performative shape and form, albeit on the periphery. The idea of voluntary material simplicity strikes a chord with the lawful forest, and reminds us of those continuities that have recurred across critical time and space, where a 'whole world of comparable instances' opens up (again). It is not unremarkable that many of these performances took place in or nearby the forest, as they have countless times before. At the midpoint of the second half of twentieth-century Australia, the remnants of the Big Scrub or the wild ancient forests of Tasmania were flashpoints of disruption, forested backdrops where young idealistic actors performed new, brave and different ways of doing.

---

[146] Stickells 'Negotiating' (n 2); Stickells, 'Housing the Farmers' (n 12); *Not Quite Square: The Story of Northern Rivers Architecture* (Lismore Regional Gallery, 2013).

[147] Tragically, Lapoinya Forest was razed. Bob Brown (n 90) describes the loggers' victory as 'pyrrhic', however, given the precedent the High Court case established in 2017 for lawful forest protest.

6

A Future Dystopia

RIVER RED GUM

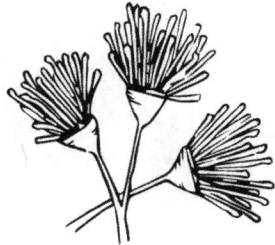

Our book dwells on a fault line between two tectonic forces: on the one hand, the seemingly inexorable 'progress' of enclosure, and on the other, the resistance to its 'rational' linear logic and its (seemingly) unstoppable trajectory. Through stories of rupture, of times when these tectonic forces collided in plainer view, *The Lawful Forest*'s critical history of the last millennia reveals *continuities*, ongoing, recurring moments of societal inflection. This is the common thread that joins our critical history of property, protest and (claims to) spatial justice.

These inflection points are profound because they bring into sharper focus the nature of our relations with land; as both a physical space, and the abstract spatial orderings that property and its laws enliven. Such relations have played out against one unbending truth, that of land's finitude. In writing of the homeless in American cities in the 1990s, Jeremy Waldron spells out what land's finitude means for 'those without property and those without community',[1] and the stark implications this has for spatial justice. It is a simple truth worth repeating: 'everything that is done has to be done somewhere. No one is free to perform an action unless there is somewhere he (sic) is free to perform it.'[2] Another quote underscores this same pressing issue of land's scarcity, but it also reveals an alternate counter-truth. This is the spatial implication of what Chapter 3 first describes as the *commonweal*, where 'the expansion of public wealth in land creates more space for everyone, while the expansion of private wealth in land reduces the space available for others'.[3]

---

[1] Jeremy Waldron, 'Community and Property – For Those Who Have Neither' (2009) 10 *Theoretical Inquiries in Law* 161.

[2] Jeremy Waldron, 'Homelessness and the Issue of Freedom' (1991) 39 *UCLA Law Review* 295, 296 ('Homelessness').

[3] George Monbiot (ed.), Robin Grey, Tom Kenny, Laurie Macfarlane, Anna Powell-Smith, Guy Shrubsole and Beth Stratford, *Land for the Many: Changing the Way Our Fundamental Asset Is Used, Owned and Governed* (Labour Party, 2019) 12.

These inflection points likewise reveal the spatial choices made over time; where the commonweal confronts private wealth, and the linear path of enclosure is briefly disrupted before the status quo resumes. Critically, they provide the briefest of glimpses into *other* paths not taken, those against-the-grain spatial alternatives that would appear as random aberrations but for a longer-term, critical view of land and its history. In the twenty-first century, land's finitude now means that our available options are increasingly fewer and more limited. We are simply running out of space – and time.

We are at another inflection point now, perhaps the gravest ever. We approach the edge of the precipice, the cusp of William Gibson's 'Jackpot', the literary device employed at the very beginning of this book. The ever-escalating climate crisis, the global pandemic, and rolling political disruption all hint (or most likely shout) of what Sophie Cunningham describes in Chapter 1 as 'the unravelling'. The context of the great pandemic of the early twenty-first century, in contrast to earlier plagues like the Black Death in Chapter 3, is that we are literally running out of space. We have destroyed the forest and as a result, the pathogen roams freely amongst us. It is well past time to critically scrutinise these *other* spatial paths, their historical lessons little learnt, and their glimpsed promises of 'shadow revolutions' that at best were only fleetingly enacted.

Given the rich history explored in the preceding chapters, how is it that property (and its resultant spatial ordering) has come to be understood through such a narrow ideological prism and, relatedly, why we are now standing on the precipice of ecological collapse? The Introduction and especially Chapter 1 propose in greater detail a number of theoretical responses to these interrogations. And, of course, poor spatial decisions are only part of this existential malaise. Yet intrinsic to our flawed relations with land and property is a deliberate, ongoing, and often violent, process of erasure. A key purpose of this book has been to unearth the inherent diversity of common law property and its rich alt-potential through a critical engagement with its history. It is also to record and lay bare the violence of this erasure, and its unequal, dispossessing effects. Yet while brutally efficient, this erasure is not total. A metaphysical *other* lingers, even if the physical 'forest' is mostly gone. In earlier chapters, we describe this lingering as akin to a haunting. Sarah Keenan reminds us *why* 'ghostly matters' matter:

> [T]o acknowledge and study ghostly matters is important in recognizing the complex ways that power operates . . . from state institutions and inescapable meta social structures such as racism and capitalism, through

[to] countless, seemingly innocuous everyday things, practices and understandings.[4]

In focusing on space and place, *The Lawful Forest* recognises land's centrality as 'the plinth of power',[5] and the unremitting manipulation of property that has privileged and entrenched the dominance of the few over the many. As Chapter 1 points out, it is difficult to discern Keenan's 'ghostly matters' given that enclosure's erasure has been so successful. The lawful forest *is* lost to plain sight. Property's diverse pasts (and presents) vanish into a 'universalising and totalising' version of property that flattens and decontextualises our relations with land. To reverse this paradigm is to engage in careful, quiet and patient observation, to 'see' again (in a Rosean sense) the significance of those 'countless, seemingly innocuous everyday things', and to understand how they may give some inkling of *other* nuanced and relational connections with space that linger out of plain sight. Luke Bennett and Antonia Layard argue similarly of the need to transform amorphous 'affective geographies of matter'[6] into more rigorous (and we would argue, patent) legal-geographic frameworks.

Beyond theory, the critical histories of *The Lawful Forest* are told in successive chapters. In the *Ancient Forest* of Chapter 2, the Forest Charter of 1217 provides an example of an early inflection point, a pushing back against royal prerogative and power in Norman England. The Charter is remarkable on many grounds, not least in its documenting of what came before, an account of a landed polyglot of diverse and pluralist property laws and customs – the so-described 'ancient forest liberties of good and lawful men'.

Chapter 3 moves forward – detailing glimpses of utopia – the ground-up legitimacy of the *commonweal* and claims to equality, and the drawn-out battle over the legitimacy of enclosure, fought in the public square, in court, and in literature. In Chapter 4, these glimpses took concrete form, albeit by way of fleeting prefigurative protests that performed working models of another way of relating to space and community. During this period, the pace of enclosure – and erasure – intensified, and by the eighteenth century the concept of private property began to be articulated and increasingly legitimised. The urban environment also began to take centre stage in this new phase of contestation over spatial ordering, and claims to the right to the city were articulated and enlivened.

---

[4] Sarah Keenan, *Subversive Property: Law and the Production of Spaces of Belonging* (Routledge, 2015) 167 (*Subversive Property*).

[5] Historian E P Thompson, cited in Nick Hayes, *The Book of Trespass: Crossing the Lines That Divide Us* (Bloomsbury, 2020) 23.

[6] Luke Bennett and Antonia Layard, 'Legal Geography: Becoming Spatial Detectives' (2015) 9(7) *Geography Compass* 406.

Finally, in Chapter 5, we return to the forest where an idealistic generation in Australia acted on an increasing awareness of the ecological damage being wrought by enclosure and its related consumptive ideology. Their moment of societal inflection was seen in a rural 'back to the land' movement, and the forging of a new community that would rise up to defend the forest from destruction.

The lessons of this selective critical history highlight the need to reanimate the *zeitgeist* of inflections past, and to channel it into the urgent imperatives of our age. For example, Tim Dunlop suggests that we could profitably 'lean into' the zeitgeist of the late 1960s/early 70s (Chapter 5) and understand why that era's idealism has descended into today's 'despair of the coronavirus world and climate change'.[7] Dunlop lays the blame squarely on neoliberalism. 'Like any good story, there is a villain, and here that villain is known as neoliberalism. This hustler came to town in the late 1970s and early 1980s, wearing a shiny suit and a shit-eating grin and promised us the world.'[8] Neoliberalism's 'greatest triumph', Dunlop argues, was 'to make us cynical about all things 60s.' It was a sleight of hand that was readily embraced:

> The transition from the '60s and '70s of hippiedom to the neoliberalism of the next 50 years was made possible because the political class—either the elites in government or those who influence government—were able to dismiss the '60s as an aberration, and in so doing foreclose on the possibility of the ideas and values that the '60s animated.[9]

Such is erasure's universality, capital's foreclosure on the pejorative *other*, and neoliberalism's hyper-intensification of 'all of the above' that our descent into the ecological abyss has only picked up speed. Yet, as this book (and Dunlop) submits, these past inflections present as 'proposal[s] for how we might reconnect'. Ominously, they may be our best last shot. These are accounts consistent with critical geography's theorisation of space and place; the dynamic, performative, co-constituted 'throwntogetherness' of countless 'heres-and-nows'[10] that reveal in their varied telling a world of striking *continuities*. Such continuities are this book's recurring themes and stories of hope: the critical histories of property, protest, and spatial justice.

---

[7] Tim Dunlop, 'A Love Letter to the Days of Future Past', *Meanjin* (Summer 2020), available at <https://meanjin.com.au/essays/a-love-letter-to-the-days-of-future-past/>.

[8] Ibid.

[9] Ibid.

[10] Doreen Massey, *For Space* (SAGE, 2005) 140.

Of Property

Central to understanding (common law) property, and its role in this book's critical history, is to acknowledge its inherent plurality. It is also to realise how comprehensively this plurality has been erased in plain sight, and to keep to front of mind the concomitant reformulation (and constant reiteration) of property through a 'single text' view of the law.

The idea that common law property is pluralistic is alien to most legal minds. For generations, property law has been taught in an a-contextual vacuum, a disparate collection of real property statutes and unanchored doctrines that portray property as a 'disembedded legal superstructure'[11] – an arcane subject avoided by students with its dull focus on persons and their legal relations about things. For example, Stuart Banner attributes the ascendance of the bundle of sticks metaphor to its uncritical adoption by property scholars in the United States in the late nineteenth and early twentieth centuries.[12] The ripple effects of earlier pedagogical choices and an ongoing lack of curiosity about property's diversity (past and present) consign the discipline to its narrow ideological bandwidth. Meanwhile its core 'values' – exclusion, alienation, commodity and individualism – persist in distorting all that enters its orbit. Yet, as this book shows, the conceptualisation of property as exclusively private is a relatively recent phenomenon, the culmination of a vast investment made by the partisan, self-interested proponents of enclosure.

Chapter 2 speaks to this ancient diversity, of what lies on the forest floor. The legal landscape of pre-Norman England was a polyglot of village-based law, customised to place and community. It was a 'coherent, stable and enduring legal order neither state-like nor stateless', an order 'best understood on its own terms'.[13] The Norman Invasion, contrary to popular mythology, did not efface this legal diversity, replacing it with a blank slate, but rather overlaid it with incremental change. The later documenting of common rights continued to reflect property's evolving diversity, Jeanette Neeson observing how 'on the ground, the range of common produce was magnificently broad,

---

[11] Nicholas Blomley, *Law, Space, and the Geographies of Power* (Guilford Press, 1994) (*Law, Space, and Power*).

[12] Stuart Banner, *American Property: A History of How, Why, and What We Own* (Harvard University Press, 2011).

[13] Tom Lambert, *Law and Order in Anglo-Saxon England* (Oxford University Press, 2017) 156. Levi Roach concurs, noting that legal historian Patrick Wormald (writing in 1977) 'realized more clearly than any before him how . . . different later Anglo-Saxon legal culture was from its modern counterpart', Levi Roach, 'Law Codes and Legal Norms in Later Anglo-Saxon England' (2013) 86 *Historical Research* 465.

the uses to which it was put were minutely varied, and the defence of local practice was determined and often successful'.[14]

Chapter 3 likewise records the robust defence of 'local [property] practice' in fourteenth- to sixteenth-century England. It notes the early resistances to enclosure through rural and urban anti-enclosure riots, and the performances of communal claims to property whereby townsfolk 'walked the bounds' of their commons in an explicitly symbolic and political act of property pluralism.

In Chapter 4, the Diggers of St George's Hill and elsewhere performed new property concepts and foundational ideologies foreign to the orthodoxies of the seventeenth century, attracting the ire of commoners and private owners alike, and exploiting the trope of the 'Norman Yoke' to assert a universalised ownership of 'vacated' Crown lands and wastes by the poor and oppressed.

Chapter 5 depicts how transplanted common property models found fertile soil in the colonial state, flourishing on the 'countercultural terroir' of north-eastern NSW, Australia. These Aquarian property experiments also remind us of how pluralism is never far from the black-letter surface, with their skilful adaptation of legal devices (such as the co-operative) to encapsulate common property norms within an external landowning entity.[15] As John Orth pithily surmises, 'all property law is local . . . the place where the land lies'.[16]

Orth's reflection on property's 'localness' speaks to another factor vital to pluralism, the overlooked, long-discredited significance of context. Pluralism remains a present reality, but we are conditioned to be blind to it because the process of decontextualisation has been so far-reaching and so total. Thomas Steinberg's 'real property isn't', or Nicole Graham's 'lawscape', amongst others, are scholarly endeavours that seek to re-centre property law to its material origins and the particularity of context.[17] Like the urban farmers and gardeners of Detroit discussed in the Introduction, performances of property plurality demonstrate how diverse tenures emanate from the ground up.

---

[14] Jeanette Neeson, *Commoners: Common Right, Enclosure and Social Change in England, 1700–1820* (Cambridge University Press, 1993) 313.

[15] John Page, 'Common Property and the Age of Aquarius' (2010) 19 *Griffith Law Review* 172.

[16] John Orth, *Reappraisals in the Law of Property* (Ashgate, 2010) vii.

[17] Another reflection on modern property plurality is Blomley's study of hybrid property, seen through a flowering bath-tub placed on a street verge in Strathcona, Vancouver, Nicholas Blomley, 'Flowers in the Bathtub: Boundary Crossings at the Public–Private Divide' (2005) 36 *Geoforum* 281.

However, the modern tale of property's plurality is mostly a depressing one of universalising erasure. To paraphrase Nicholas Blomley, what has been lost is not just the physical acreage of diverse, non-private tenures, but their alt-potential.[18] In a sense, this latter loss is the most profound, the 'unknowing' of other ways of relating to space that threaten to foreclose on our earthly future.

As this book shows, the erasure of property plurality was not the outcome of a single supervening act or moment, but rather a gradual process of enclosure that gathered pace and intensity over centuries. Even the Norman Invasion did not erase pluralism, although it did enlarge and institutionalise feudalism. However, in the scheme of things, Norman feudalism did not endure long. It was dramatically undercut by the freedoms of alienation introduced by *Quia Emptores* in 1235, and feudal incidents thereafter started to wither on the vine. By 1660, it was consigned to (near) antiquity by the Tenures Abolition Act.

The Norman Invasion, however, did provide a template for the invasion of Australia and its more successful erasure of Indigenous property rights (and, indeed, Indigenous peoples themselves) through both violence and the fiction of *terra nullius*. The doctrine of tenure found a new life under the harsh Australian sun, such that the continent was seen as a fortuitous *tabula rasa*, a blank slate wiped clean and made ripe for land theft and white colonisation.[19] Analogies have also been drawn between what was then the contemporaneous erasure of common rights in the metropole, and Indigenous erasure in the colony.[20] It is only in recent times that this myth is being properly acknowledged and unpacked – which has led to the unearthing of the law of the land (or Country). The *Milirrpum* case of 1971 (Chapter 5) not only failed to 'see' Indigenous relationships to land as propertied, but also failed to give due consideration to the common law's own history of pluralism, a double omission which makes the former's erasure even more egregious, if that is possible.[21]

---

[18] Blomley, *Law, Space, and Power* (n 11).

[19] A comparative study of British colonial erasure, including the colony of New South Wales, can be found in Stuart Banner, *Possessing the Pacific* (Yale University Press, 2008). And as Chapter 1 details, the orthodox conception of space renders it neutral and inert, a *tabula rasa*. 'Legal judgments, executive powers, legislation and legal commentaries tend to treat space as something to be planned over, built on, cultivated, bought, sold and/or protected; a blank canvas . . . to be smoothly acted upon', Keenan, *Subversive Property* (n 4) 21.

[20] Andrew Buck, *The Making of Australian Property Law* (Federation Books, 2006).

[21] *Milirrpum v Nabalco Pty Ltd and the Commonwealth of Australia* (1971) 17 FLR 141 (Blackburn J).

In colonial Australia, the Torrens title system, described as 'a system of title *by* registration, not a system of registration of title',[22] has taken erasure to the next level. Devised in the mid-nineteenth century and exported offshore in the years that followed, the Torrens system erased historic common law claims that were based on pluralist (and often competing) property concepts.[23] For example, performative doctrines such as prescription, or place-based rituals such as walking the bounds, were delegitimised, dismissed as 'piling fiction upon fiction'.[24] Torrens title likewise adopts the 'improvement' playbook, promising the *certainty* of indefeasible title and the 'blank slate' of the one *authoritative* title deed. As its founder Robert Torrens explained (in words reminiscent of Blomley), 'the entire fabric must be razed to the ground and a new super-structure substituted . . . he would take a sponge and rub the whole out'.[25]

Torrens title was also particularly effective in erasing Indigenous land claims, with the grant of a registered fee simple the ultimate 'rubbing out' of native title.[26] Deirdre Howard-Wagner argues that through the Northern Territory Intervention (now Stronger Futures), 'the law plays a significant role in the institutionalization of settler-colonial norms' particularly in relation to land by targeting Indigenous communal land arrangements and title and imposing a system of individual property rights.[27] Keenan describes Torrens title as an exercise in 'radical temporal dislocation'[28] where the only 'time-traveller' permitted to operate the machine is a white male, and Indigenous claims to land are incapable of being reflected in its mirror-like, self-fulfilling register.[29]

---

[22] Barwick CJ in *Breskvar v Wall* (1971) 126 CLR 376 (emphasis added).

[23] Keenan critiques the deeds-based Old System title as 'a social and physical world ordered by class and inheritance', Sarah Keenan, 'From Historical Chains to Derivative Futures: Title Registries as Time Machines' (2019) 20(3) *Social & Cultural Geography* 283, 287 ('Time Machines').

[24] *Williams v STA* (2004) 60 NSWLR 286.

[25] Robert Torrens, cited in Peter Butt, *Land Law* (Thomson Reuters, 6th ed., 2010) 744. Judges likewise demonised the Old System, describing its complex documentation as 'difficult to read, disgusting to touch and impossible to understand', ibid.

[26] *Fejo v Northern Territory* [1998] HCA 58.

[27] Deirdre Howard-Wagner, 'Reclaiming the Northern Territory as a Settler-Colonial Space' (2012) 37/38 *Arena Journal* 220, 240, citing Patrick Wolfe, 'Settler Colonialism and the Elimination of the Native' (2006) 8(4) *Journal of Genocide Research* 387, 388. Howard-Wagner does caution, however, that despite the best efforts of the 'settler-colonial endeavour . . . the traditional owners of the land have managed to successfully resist the erasing features of settler colonialism and maintain connections with their land', ibid 240.

[28] Ibid 291.

[29] Keenan, 'Time Machines' (n 23) 283.

Digitisation has sent this process of erasure into hyper-drive. Electronic conveyancing and real-time registration means that Keenan's equating of Torrens land as 'an almost liquid asset, a feat previously thought impossible due to land being an immovable, limited, unique and necessarily shared resource',[30] is given new, vastly 'improved' reductive impetus. With its increasing efficiency gains, people can be unhoused by digital investors at the stroke of a computer key, and ancient human performances are further lost to purely online administrative transactions. 'The world external to the [Torrens] registry can blur away into insignificance'[31] in what is an erasure and decontextualisation of time and space *par excellence*.

Finally, a closing discussion of erasure would not be complete without reference to the power of words. Language has always been a key tool of enclosure – words are effective at constructing reality – such that we can no longer see what is in plain sight. As Blomley observes, if certain spaces 'do not look like property to us, we have tended to ignore them'.[32] Terms such as *improvement, efficiency, progress* or *particularise* are technicalities that hide a brutal history, euphemisms for enclosure's often-violent dispossession of the *other*. As Chapter 1 argues, legal pluralism is a threat to the private hegemony, such that words that remind us of its extant state are dismissed as 'not propertied' or 'illegitimate' in sanctioned property discourse.

Protest

Our critical history of protest is chiefly a record of resistance to spatial enclosure, from the physicality of the rural commons to the meta-physicality of the right to the city. Yet what lies beneath, the subliminal sub-text to the bright lines of anti-enclosure protest, is a more threshold resistance to social control – the lower-level, repressive violence that goes into maintaining the status quo.

What is telling from our critical history of protest is that often protests were demands that the law be upheld, not defied. The pannage suit at Leigh Sinton in 825 (Chapter 2) did not seek to overturn the 'law' but rather sought enforcement of long-standing custom as to the extent of the community's rights to its swine pastures. Similarly, in 2016, Bob Brown and Jessica Hoyt were arrested for protesting against the unconstitutional proscription of protest in Tasmanian forests (Chapter 5). And of course, enclosure was originally a breach of the common law, such that anti-enclosure riots, early

---

[30] Ibid 285.
[31] Ibid 292.
[32] Nicholas Blomley, *Unsettling the City: Urban Land and the Politics of Property* (Routledge, 2004) 8.

legal suits, and practical actions such as the breaking of hedges, were made in furtherance of the law, not against it (Chapters 3 and 4). Yet where the law is administered and enforced by law-breakers, the imperatives of protest shift over time from adherence to defiance of unjust laws.

In Chapter 2, outrage against the Norman forest law was incandescent, with forest protest successfully pushing back against royal enclosure for over 750 years through the remarkable Forest Charter. In Chapter 3, the Peasants' Revolt, later urban riots (that developed the notion of the *commonweal* as both communal property and a stake in local governance), and the literary protest of Thomas More's *Utopia*, were all unified in their resistance to enclosure. In Chapter 4, protest became more overtly political, as exercises in early prefigurative politics, and brief enactments of concrete or working utopias. The *campyng tyme* or *commotions* of 1549 saw large-scale utopian protest camps spring up in defence of the commons. These times also presaged increasingly violent patterns of repression and erasure. From the Diggers' proto-communist landed experiments to the Paris Commune (and its notion of *communal luxury* echoing the earlier politics of the commonweal), and the *manifestation* of Lefebvre's right to the city in Paris 1968, this chapter illustrates how a 'pushing back' against enclosure took multiple forms, material and intellectual. Finally, the rural communes and 'hippy' ideals of the Age of Aquarius (Chapter 5) were wellsprings of forest protest in modern Australia, protests that close a loop of sorts by reminding us of latent forest liberties and ancient traditions of defending the lawful forest in the public square.

Across time and space, public or communitarian space was and remains a flashpoint and *situs* of protest. After all, this was the terrain, physical and abstract, that enclosure sought to extinguish. During the COVID pandemic, we have, for example, seen the criminalisation of protest, with public health orders being used as a convenient cover to outlaw protests in public space.[33] Police in the United States have been accused of exploiting public health curfews to criminalise and violently harass people protesting police brutality as Black Lives Matters (BLM) protests spread into the streets.[34] In the

---

[33] Louise Boon-Kuo, Alec Brodie, Jennifer Keene-McCann, Vicki Sentas and Leanne Weber, 'Policing Biosecurity: Police Enforcement of Special Measures in New South Wales and Victoria during the COVID-19 Pandemic' (2020) *Current Issues in Criminal Justice* 6–7, DOI: 10.1080/10345329.2020.1850144.

[34] Human Rights Watch, '"Kettling" Protesters in the Bronx: Systemic Police Brutality and Its Costs in the United States' (30 September 2020), available at <https://www.hrw.org/report/2020/09/30/kettling-protesters-bronx/systemic-police-brutality-and-its-costs-united-states>; Christopher Petrella, 'How Curfews Have Historically Been Used to Restrict the Physical and Political Movements of Black People in the U.S.', *The Washington Post* (4 June 2020), available at <https://www.washingtonpost.com/nation/2020/06/03/how-curfews

Australian state of NSW, public health orders were also used to repeatedly shut down BLM protests planned for city streets and parks.[35]

One of the greatest ironies of our critical history of protest is the fate of St George's Hill in Surrey; the former Crown lands and wastes where the original Diggers camp of 1649 momentarily took life. Today the land is 'a gated enclave of London's rich and famous, a place of exclusivity, of private golf courses, tennis courts and mansions on multi-acre lots'.[36] In 2019, activist/blogger Guy Shrubsole writes of breaching the perimeter fences of St George's Hill, spotting a break in the chain-link to enter and make his own small gesture of protest, 'treading in the footsteps of Winstanley, trespassing upon land once held in common but now held offshore'.[37] Shrubsole notes the 'cruel joke' that is St George's Hill, the 'supreme irony of history', how a once 'radical communal experiment came to be completely enclosed . . . '.[38] That Shrubsole is forced to make his (illegal) gesture of protest surreptitiously and alone; 'digging up the earth, sinking my fingers in the leaf mould and the thick black loam, smelling its rich peaty scent',[39] exemplifies the fiercely anti-protest nature of private property, its a-contextuality, and what happens when protest fails to shift the linear progress of enclosure.[40] The cruel irony that is St George's Hill also leads into the third of our overarching themes: that of spatial justice, and how private property's 'cult of exclusion,'[41] and its associated landed politics of theft, is core to spatial (in)equity.

-have-historically-been-used-restrict-physical-political-movements-black-people-us/>; Fernanda Echavarri, 'Curfews "Definitely Work" – If We Define "Work" as Increasing the Possibilities for Police Violence', *Mother Jones* (2 June 2020), available at <https://www.mot herjones.com/crime-justice/2020/06/do-curfews-work-confusion-los-angeles>.

[35] See, e.g., *Commissioner of Police v Gray* [2020] NSWSC 867 (4 July 2020); *Commissioner of Police (NSW) v Gibson* [2020] NSWSC 953 (26 July 2020); *Raul Bassi v Commissioner of Police (NSW)* [2020] NSWCA 109 (9 June 2020). They also served to criminalise university students attempting to protest the Federal Government's fee hikes and austerity-driven restructuring of the higher education sector.

[36] Thomas Willard, 'The Free Enjoyment of the Earth: Gerrard Winstanley on Land Reform', in A Classen (ed.) *Rural Space in the Middle Ages and Early Modern Age: The Spatial Turn in Premodern Studies* (De Gruyter, 2012) 888.

[37] Guy Shrubsole, *Who Owns England? How We Lost Our Green & Pleasant Land & How to Take It Back* (William Collins, 2019) 213.

[38] Ibid 212.

[39] Ibid 213.

[40] John Page, *Public Property, Law and Society: Owning, Belonging, Connecting in the Public Realm* (Routledge, 2021).

[41] Hayes (n 5).

And Spatial Justice

The critical history of spatial justice explored in this book inevitably circles back to an early statement we made in Chapter 1: 'Here, [land's] finitude by necessity implicates spatial justice. It is land's *scarcity* and the grievance over its uneven distribution that lies deeply unresolved on the forest floor.'

This uneven distribution plays out in different ways. It can mean, in its crudest sense, the gross injustice of having little if any physical space *to do* and *to be*, which is Waldron's point. The physical enclosure of space within private fences and its universal dispossession of the many is (after all) the inexorable end-logic of enclosure. In pandemic London, such *spatial* inequality is depicted in the harrowing scenes of Londoners 'swelter[ing] in tiny flats, or edg[ing] round each other in miniscule parks, desperate for a sense of space and freedom'[42] (see the Introduction). Transfer that imagery from the global north to cities such as New Delhi, and the tragedy of spatial inequity is only magnified in its tragic consequences.

Spatial inequality is also a legacy of power imbalances, of structural racism, colonialism, or the effects of capital. Notions of who *belongs*[43] in public spaces have been explored through the prism of race in the United States, from Detroit's Liberated Farms to a Black birdwatcher in New York's Central Park and the spatial premise of BLM protests. Colonialism and racism have likewise combined with potent force to enact spatial injustice in the colonial state. In Australia, the (ongoing) fiction of *terra nullius* ignores Indigenous relationships with land, such that the continent presents as a blank slate ripe for colonisation, alienation and Indigenous dispossession. Finally, the rise of capitalism has baked in another manifestation of spatial inequity, first by rural dispossession creating a class of landless labour, and then by the increasing curtailment of freedom of movement. Or where the modern-day shift from the public square to private (pedestrian) malls transforms citizens into consumers, whose only right is consumption, given that protest, freedom of expression, or freedom of movement is antithetical to commercial pursuits.[44]

---

[42] George Monbiot, 'Lockdown Is Nothing New: We've Been Kept Off the Land for Centuries', *The Guardian* (22 April 2020), available at <https://www.theguardian.com/com mentisfree/2020/apr/22/lockdown-coronavirus-crisis-right-to-roam>.

[43] Belonging is 'the state of fitting smoothly, or without trouble, into a conceptual category or a material position', Keenan, *Subversive Property* (n 4) 12.

[44] See, e.g., Katharine Gelber, 'Pedestrian Malls and Local Government Powers: Political Speech at Risk' (2003) 5 *UTS Law Review* 48; Michael Head, 'The High Court Further Erodes Free Speech' (2013) 38(3) *Alternative Law Journal* 147; and the cases of *Coleman v*

In Chapter 2, we see an early example of spatial injustice through the dispossession of common rights in afforested lands, which at its apex denied commoners' access to forest commons, their livelihoods, and ancient sites of sociable festivity. In Chapter 3, the end of the manorial system and claims to freedom of movement and labour power were met with the Statute of Labourers. This blunt legislative instrument drew the powerful link between the control of labour power and property. It was also an example of the use of elite power gained through landholding to enact social control.

Chapter 3 also records early anti-enclosure riots, where 'wool was king' and the overriding priority was economic. The imperative to fence sheep pastures (and the denial of access to the commons this caused) enacted an early wave of rural spatial injustice. The wool industry also led to the commercialisation of agriculture and a shift in land use and labour practice. As Hythloday observed of sheep in *Utopia*:

> [I]n whatever parts of the land the sheep yield the softest and most expensive wool, there the nobility and gentry, yes, and even some abbots though otherwise holy men, are not content with the old rents that the land yielded to their predecessors . . . For they leave no land free for the plough: they enclose every acre for pasture; they destroy houses and abolish towns, keeping only the churches, and those for sheep-barns . . .[45]

Wool likewise played an over-sized role in Indigenous dispossession in parts of colonial Australia, where vast squatting 'runs' for sheep grazing were stolen from Indigenous owners, creating a 'bunyip aristocracy' that mimicked the English upper class in their manorial estates and penchant for spatial injustice.

Kett's Rebellion in Chapter 4 likewise focuses on spatial inequity. A key driver of popular dissent was the rise of a landless class – an early example of commodified land driving people to destitution – a development with potent parallels to modern-day urban homelessness[46] and the architecture of

---

Sellars (2001) 181 ALR 120; *Attorney-General (SA)* v *Corporation of the City of Adelaide* [2013] HCA 3.

[45] Thomas More, *Utopia* (1516), ed. R M Adams (W.W. Norton, 1992) 12, cited in William Gerard and Eric Sterling, 'Sir Thomas More's Utopia and the Transformation of England from Absolute Monarchy to Egalitarian Society' (2005) 8(1) *Contemporary Justice Review* 75, 78. Interestingly, the early keeping of pigs, unlike sheep, did not require such an emphasis on fences and bounded-ness. Swineherds were common in 'glandiferous woodland' (Chapter 2) and pannage was a curious urban practice in eighteenth- and early nineteenth-century New York, Hartog, 'Pigs and Positivism' (1985) *Wisconsin Law Review* 899.

[46] Waldron, 'Homelessness' (n 2) 296.

Vagrancy Acts, move-on laws, and hostile architectural design that perpetu-
ates this particular version of spatial injustice.[47]

In the early twenty-first century, the COVID-19 pandemic has highlighted and exacerbated underlying issues of spatial injustice, with significant
evidence of the discriminatory and selective enforcement of public health
orders against specific population groups.[48] These discriminatory enforcement practices merely build on a long history of over-policing of particular
groups in society, often through the selective use of discretionary 'stop and
search' or 'move on' powers and laws prohibiting loitering, offensive language or public drunkenness.[49] In this way, policing serves to regulate who is
permitted to occupy public space, and under what (and whose) condition(s).
In Australia, these practices have particularly targeted Aboriginal and Torres
Strait Islander peoples, in addition to culturally and linguistically diverse
communities, and people experiencing homelessness and mental health
issues.[50] In the United States, the violent use of these powers against Black
people has been extensively documented.[51] Indeed, the BLM protests in both
countries are a direct result of these discriminatory policing practices and
their deadly results.[52]

[47] The phenomenon of 'hostile architecture' is designed to deny homeless access to public space – rendering benches or areas under eaves unable to be slept on.

[48] Osman Faruqi, 'Compliance Fines under the Microscope', *The Saturday Paper* (18 April 2020), available at <https://www.thesaturdaypaper.com.au/news/health/2020/04/18/compliance-fines-under-the-microscope/15871320009710>; Boon-Kuo et al. (n 33) 4–6; Covid Policing, 'COVID Policing Weekly Roundups' (2020), available at <https://covidpolicing.org.au/>.

[49] Boon-Kuo et al. (n 33); Law Council of Australia, 'Broader Justice System Players', in *The Justice Project: Final Report* (Law Council of Australia, 2018) 24–30.

[50] Daniel Haile-Michael and Maki Issa, '*The More Things Change, The More They Stay the Same*': Report of the FKCLC Peer Advocacy Outreach Project on Racial Profiling across Melbourne (Flemington and Kensington Community Legal Centre, 2015); Law Council of Australia (n 49) 24–30.

[51] John Wihbey and Leighton Walter Kille, 'Excessive or Reasonable Force by Police? Research on Law Enforcement and Racial Conflict', *The Journalist's Resource*, Harvard Kennedy School Shorenstein Centre, available at <https://journalistsresource.org/studies/government/criminal-justice/police-reasonable-force-brutality-race-research-review-statistics/>; Bernadette R Hadden, Willie Tolliver, Fabienne Snowden and Robyn Brown-Manning, 'An Authentic Discourse: Recentering Race and Racism as Factors that Contribute to Police Violence against Unarmed Black or African American Men' (2016) 26(3–4) *Journal of Human Behavior in the Social Environment* 336.

[52] Alicia Garza, 'A Herstory of the #BlackLivesMatter Movement', *The Feminist Wire* (6 December 2014), available at <http://www.thefeministwire.com/2014/10/blacklivesmatter-2/>; Yamiche Alcindor, 'Black Lives Matter Coalition Makes Demands as Campaign Heats Up', *The New York Times* (1 August 2016), available at <http://www.nytimes.com

Many studies of these discriminatory and violent policing practices have highlighted their political dimensions. For example, David Jacobs and Robert O'Brien found (in 1998) 'that the police use of lethal force varies with the degree of inequality between the races, the presence of blacks, and local political arrangements that increase black control over the behavior of law enforcement personnel'.[53] They argue that '[s]uch results are consistent with claims that state violence is used in racially unequal jurisdictions to preserve the existing order.'[54] John Logan and Deidre Oakley similarly argue that 'a routine function of policing' is the protection of 'the mainstream United States from the perceived risk from its "ghetto" under-belly'.[55] They explain that policing is just one element of 'a system of separation and containment' – a system that also includes 'residential segregation, inequality in public schools, displacement, as well as health and environmental inequality'.[56] In their view, 'all of these forms of spatial containment' can be described as 'ghettoization' – due to the history of ghettos as a mechanism for 'containment and control' even when they are spatially unbound (or 'iconic' in nature).[57]

Writing about the environmental inequality of this system of spatial containment, David Pellow argues that 'if we think of environmental racism . . . as a form of authoritarian control over bodies, space, and knowledge systems[,] then we can more effectively theorize it as a form of state violence'.[58] This link

/2016/08/02/us/politics/black-lives-matter-campaign.html>; Professor Marcia Langton, 'Why the Black Lives Matter Protests Must Continue: An Urgent Appeal by Marcia Langton', *The Conversation* (5 August 2020), available at <https://theconversation.com /why-the-black-lives-matter-protests-must-continue-an-urgent-appeal-by-marcia-langton -143914>. However, the US movement, focused as it is on people of African descent, places inadequate attention on the significance of land and sovereignty for Indigenous peoples, Chelsea Bond, 'We just Black Matter: Australia's Indifference to Aboriginal Lives and Land', *The Conversation* (16 October 2017), available at <https://theconversation.com/we -just-black-matter-australias-indifference-to-aboriginal-lives-and-land-85168>.

[53] David Jacobs and Robert M O'Brien, 'The Determinants of Deadly Force: A Structural Analysis of Police Violence' (1998) 103(4) *American Journal of Sociology* 837.

[54] Ibid.

[55] John R Logan and Deirdre Oakley, 'Black Lives and Policing: The Larger Context of Ghettoization' (2017) 39(8) *Journal of Urban Affairs* 1031. Slave ownership was, of course, a key factor in the foundational development of the police force in the United States (and the Second Amendment) to control slave uprisings and protect white people's 'property rights'.

[56] Ibid 1032.

[57] Ibid.

[58] David Pellow, 'Toward a Critical Environmental Justice Studies: Black Lives Matter as an environmental justice Challenge' (2016) 13(2) *Du Bois Review* 221, 233.

between social and environmental justice clearly relates to other examples of governments using the pandemic as cover to introduce controversial policies. For example, the Australian government has used the COVID-19 economic downturn as an excuse to push for a 'gas-fired recovery' despite clear evidence that the world is moving away from fossil fuels, and indeed must do so in order to avoid catastrophic climate change,[59] in a classic example of Naomi Klein's 'Shock Doctrine' (see the Introduction).[60]

The lockdown orders imposed in response to the COVID-19 pandemic also highlighted and exacerbated *bounded* issues of spatial injustice, cases where geography matters. In Melbourne, Australia, where residents were restricted to a 5-kilometre zone around their homes, some people's newly bounded world included lush green parks, beaches and a variety of fresh food options, while others were confined to concrete grey 'geographies of nowhere' and urban food deserts. Worse still was the fate of the residents of several public housing towers in Melbourne's north, who were subjected to a police-led mass quarantine that confined them to their overcrowded high-rise apartments with no notice.[61] As Louise Boo-Kuo et al. argue, '[t]hese "hard lockdowns" involved high-visibility policing at odds with centring the health needs of residents.'[62] Indeed, '[i]t is hard to avoid the implication that these "hard lockdowns" involved racial and class markers for health risks.'[63]

While these lockdowns were an exception to the norm, the spatial inequalities that they highlighted are real and ongoing. The concrete grey 'geographies of nowhere' are already becoming urban heat islands and will only become worse as global heating continues apace. These issues of environmental justice also spread well beyond tree coverage and include air and water quality. For example, low- to middle-income households experience 90 per cent of all air pollution in Australia, while those with the highest income experience just 0.1 per cent.[64] Meanwhile, water quality issues also affect

59 The Hon. Scott Morrison, 'Gas-Fired Recovery', Media Release – Prime Minister of Australia, Minister for Energy and Emissions Reduction, Minister for Resources, Water and Northern Australia (15 September 2020), available at <https://www.pm.gov.au/media/gas-fired-recovery>.
60 Naomi Klein, *The Shock Doctrine: The Rise of Disaster Capitalism* (Penguin, 2007).
61 Department of Health and Human Services Victoria (DHHSV), *Detention Directions* (4 July 2020); Hiba Shanino, 'We Knew about Covid-19 in the Towers and Were Taking Care, but Instead of Support There Was Only Police', *The Guardian* (5 July 2020), available at <https://www.theguardian.com/commentisfree/2020/jul/05/we-knew-about-covid-in-the-towers-and-were-taking-care-but-when-we-needed-support-there-were-only-police>.
62 Boon-Kuo et al. (n 33) 8.
63 Ibid 9.
64 Australian Conservation Foundation, *The Dirty Truth: Australia's Most Polluted Postcodes*

millions around the globe – as exemplified by communities as disparate as Walgett, NSW, and Flint, Michigan.[65] The lead poisoning of the entire City of Flint made news around the world after the Emergency Manager approved a 2014 switch to the heavily polluted Flint River in order to save money to fund the construction of a private pipeline for the benefit of the wealthy white suburbs that surround the city.[66] Astonishingly, in the years since the switch, the situation has yet to be fully resolved, and residents have been left with very little in terms of justice or compensation.[67]

Whether measured in crude acreage, access to public amenities, or in the implications of racism, environmental racism, colonialism, or the distributive inequities of capital, our critical history of spatial justice highlights how the contours of private property match the geographies of spatial injustice. As David Delaney writes of critical geography and its unmasking of power relationships (in Chapter 1), our ambition is similar in its scope: 'Critical geography is indispensable for revealing the workings of power that conventional spatial imaginaries obscure and for identifying the whys, hows and wheres of injustice that are otherwise invisibilized and legitimized.'[68]

Stepping Back from the Brink

This book ends where it began, with the spectre of planetary environmental collapse, and the prophetic sense of impending doom that led the fiction writer William Gibson to temporarily abandon his work. Enclosure and dispossession continues to march to the inevitable beat of its drum – from the swine pastures of ninth-century Leigh Sinton to the climate emergency of the twenty-first. To make matters worse, the hyper-commodification of

(16 November 2018), available at <https://www.documentcloud.org/documents/5115779-ACF-Pollution-Report.html>.

[65] Lorena Allam, 'Walgett's Water Crisis: NSW Considers Options after "Concerning" Sodium Levels Found', *The Guardian* (22 January 2019), available at <https://www.theguardian.com/australia-news/2019/jan/22/walgetts-water-crisis-nsw-considers-options-after-concerning-sodium-levels-found>; Cristy Clark, 'Race, Austerity and Water Justice in the United States: Fighting for the Human Right to Water in Detroit and Flint, Michigan', in Farhana Sultana and Alex Loftus (eds) *Water Politics: Governance, Justice and the Right to Water* (Routledge, 2019) 175–88.

[66] Clark (n 65).

[67] See John Flesher, 'Flint Officials Say Progress Made toward Ending Water Crisis', *PBS News Hour* (7 December 2020), available at <https://www.pbs.org/newshour/health/flint-officials-say-progress-made-toward-ending-water-crisis>. The 'Build America Back' 2021 initiative of the Biden Administration also earmarks funding for the removal of lead pipes in certain urban water systems.

[68] David Delaney, 'Legal Geography II: Discerning Injustice' (2016) 40(2) *Progress in Human Geography* 267, 272.

neoliberalism has, over the last forty years or so, exacerbated extreme inequality and poverty – and powered the climate crisis to new lows. There is nothing (except a liveable planet) left to enclose in what looms as the ultimate act of dispossession.

So how do we step back from this dire brink? One way is to recognise, and to learn the lessons of these *other* ways of relating to land and space – what we call the places of the lawful forest. These are places replete in *communal luxury*, an ideal espoused in early articulations of the *common-weal* and given explicit name in the Paris Commune. Communal luxury could be described in other ways too, for example the concept of 'public-wealth', the land policy that the British Labour Party took (unsuccessfully) to the 2019 general election. Titled 'Land for the Many', this manifesto asserted that public wealth in land 'creates more space for everyone' and would be guided by principles of 'private sufficiency and public luxury'.[69] Howsoever described, these spatial imaginings and performances of *other* ways of relating to land provide a template, an envisioning of a way out of our present dead end.

So, what do these spatial templates look like? In collating and describing selected moments of past spatial disruption, our goal has been to provide a measure of guidance and context through example. Of course, there are numerous other precedents and approaches. Tim Dunlop suggests (from earlier in this chapter) that we need to 'lean more' into the zeitgeist of the late 1960s/early 1970s. We need to take greater heed of its paradigm-shifting moments, like the themes of Rachel Carson's *Silent Spring*, the era's environmental (re)awakenings, its ecological communes and forest protest culture (Chapter 5). Davina Cooper's 'everyday utopias' (Chapter 1) are likewise concrete, working examples of landed relations thought anew, richly imbued with their prefigurative potential to transform progressive society and politics from the ground up. Or, those enduring intentional communities whose stories the practicalities of time and space, or pandemic lockdown, have regrettably precluded. Like the Celo Community in the Appalachian Mountains of North Carolina in the United States, a 1,200-acre community of common lands established eighty years ago by the New Deal-era director of the Tennessee Valley Authority, Arthur Morgan.[70]

To describe the lawful forest is to accept its innate incommensurability, a simultaneity-of-stories-so-far that dwells in the physical and metaphysical, the grounded and the utopian. To expand *public-wealth* or *communal luxury*, property must reject the excesses of its private trappings; the hubris and

---

[69] Monbiot et al. (n 3) 12.

[70] Joshua Lockyer, *Seeing Like a Commons: Eighty Years of Intentional Community Building and Commons Stewardship in Celo, North Carolina* (Rowman & Littlefield, 2021).

artifice of exclusion, commodification and alienation. Looking backwards to prefigure a hopeful future, the template is there for all of us to see, if only we lift our line of sight. In this ancient yet modern iteration, property reflects its relational and pluralist origins, laws true to context, and in the colonial state, laws willing to learn from Indigenous worldviews. Of course, we do not have to uproot our lives and become modern-day Diggers or 1970s hippies to follow these alternate spatial paths. What is required, however, is a critical reappraisal of the dominant spatial hegemony and its structural inequities, and an openness to take greater heed of those analogous contexts and *continuities* from which the lawful forest springs.

Index

EU representative:
Easy Access System Europe
Mustamäe tee 50, 10621 Tallinn, Estonia
Gpsr.requests@easproject.com

www.ingramcontent.com/pod-product-compliance
Lightning Source LLC
Chambersburg PA
CBHW061154220326
41599CB00025B/4482